W0235183

Advances in
Nuclear Science
and Technology

VOLUME 12

Advances in Nuclear Science and Technology

Series Editors

Jeffery Lewins
University of London, London, England

Martin Becker
Rensselaer Polytechnic Institute, Troy, New York, U.S.A.

Editorial Board

Eugene P. Wigner, *Honorary President*
R. W. Albrecht
J. Gordon Balfour
V. S. Crocker
F. R. Farmer
Paul Greebler
Ernest J. Henley
Norman Hilberry
K. Oshima
A. Sesonske
H. B. Smets
K. Wirtz
C. P. L. Zaleski

A Continuation Order Plan is available for this series. A continuation order will bring delivery of each new volume immediately upon publication. Volumes are billed only upon actual shipment. For further information please contact the publisher.

Advances in
Nuclear Science
and Technology

VOLUME 12

Edited by

Jeffery Lewins
University of London
London, England

and

Martin Becker
Rensselaer Polytechnic Institute
Troy, New York, U.S.A.

PLENUM PRESS · NEW YORK AND LONDON

The Library of Congress cataloged the first volume of this title as follows:

Advances in nuclear science and technology. v. 1—
  1962—
New York, Academic Press.
  v. Illus., diagrs. 24 cm. annual.
  Editors: 1962—   E. J. Henley and H. Kouts.
  1. Nuclear engineering—Yearbooks. 2. Nuclear physics—Yearbooks.
  I. Henley, Ernest J., ed. II. Kouts, Herbert J., 1919-   ed.
  TK9001.A3   621.48058   62-13039

Library of Congress Catalog Card Number 62-13039

ISBN 978-1-4613-9918-6   ISBN 978-1-4613-9916-2 (eBook)
DOI 10.1007/978-1-4613-9916-2

©1980 Plenum Press, New York
Softcover reprint of the hardcover 1st edition 1980

A Division of Plenum Publishing Corporation
227 West 17th Street, New York, N.Y., 10011

All rights reserved

No part of this book may be reproduced, stored in a retrieval system, or transmitted, in any form or by any means, electronic, mechanical, photocopying, microfilming, recording, or otherwise, without written permission from the Publisher

Preface

The Editors take pleasure in presenting a further volume in their Annual Review Series. The present volume contains six papers that may be said to span from the theory of design to the practice of operation of modern nuclear power stations, therefore concentrating on nuclear energy as a source of electrical power. Starting with the most mathematical, and proceeding in the direction of technology, we have the Chudley and Brough account of a new interpretation of (linear) Boltzmann transport theory in terms of the characteristic or ray approach. This seems to be new in application here, but of course the method is the child of many classical studies in the solution of partial differential equations and proves to remarkably well-suited to modern computers and their numerical bases.

We might put the article by Dickson and Doncals on the design of heterogeneous cores next, with its significance for fast reactors of the future. The various "central worth" discrepancies, with their implication for safety and reliability founded on, *inter alia,* the Doppler effect, have made this a major area for resolution: to see that we can develop design methods and codes that will reconcile theory and experiment to the point at which theoretical designs could be accepted for building without the need for a full-scale mockup, as had to be done in the 1950's for the light water reactors. Al Henry has advanced the claim of thermal reactor theory to such a commanding position; we can hope that contributions such as these will advance the claims of fast reactor theory.

The gas-cooled reactors of this earlier era such as the Magnox, built in the United Kingdom, have been overtaken now by the Advanced Gas-cooled Reactors (AGRs) in a construction that has very hot, very dense and high-speed carbon dioxide coolant (sometimes described as being like red hot treacle falling over Niagara Falls), bringing substantial

technical problems. How these problems of insulation, involving detailed and meticulous welding of stainless steel
laminations, have been solved by design, is the subject of
our third paper from Furber and Davidson of the United
Kingdom Nuclear Power Company.

A fourth paper reverts to the United States for experience with the practical operating problems of Light Water
Reactors (LWRs) and their observed availability. The cost
of all reactors is high; it is not always appreciated that
fuel costs of non-nuclear alternatives is so high that the
down-time *replacement* cost of electricity for a non-working
reactor is also high. It remains a major objective, therefore, of reactor-operating utilities to improve the availability from any reactor, and American experience reported
by Burns, with PWR and BWR, will be followed with interest
worldwide.

Looking further ahead, Harms, writing from Canada, looks
to the medium term use of fuels via the variations on the
spallation reactor concepts, providing us with a scientific
overview of the available choices. It has been said of
fusion reactors, should they ever be practicable, that their
immediate worth will be countable in the number of *neutrons*
they provide. The spallation systems are based more eviden-
tly on the concept of breeding and the improved yield of
fissile fuels this provides while being recognizably feasible technology.

Cronenberg returns us to the scientific problem reviewed
once before in an earlier volume of the series, the problem
of vapor explosions with particular reference perhaps to
liquid metal coolants and their role in the future Fast Reactors. The paper illustrates well the need and the provision of "in depth" research into fundamental phenomena
that must accompany any satisfactory development and application of modern technology. Equally well, then, we need
to seek a partnership between industry and the universities
over the provision of good and useful research. We hope the
reports made available in this series contribute to that aim.

Finally, we note that Ernest Henley has retired from
the immediate role of Editor, but has been persuaded to
stay on the Editorial Board (whose other distinguished
members are the source of many suggestions and valued

criticism to authors and editors alike). Ernest Henley's first volume appeared in 1962, and so his work as an editor may be taken to span two decades. In its 11 volumes, it provides no small measure of his contribution and worth.

                              Jeffery Lewins
                              Martin Becker

Contents

CHARACTERISTIC RAY SOLUTIONS OF THE TRANSPORT EQUATION

M. D. Brough
London School of Hygiene and Tropical Medicine

and
C. T. Chudley
Bristol Polytechnic

I.   INTRODUCTION

With the onset of increasingly sophisticated reactor designs in the late 1950's, the problem of explaining the behavior of thermal neutron physics in reactors became more acute. This behavior is governed by the Maxwell-Boltzmann equation, which in its time-independent form is:

$$(\underline{\Omega} \cdot \nabla + \Sigma_s(E,\underline{r}) + \Sigma_a(E,r)) \; \phi(\underline{r},E,\underline{\Omega})$$

$$= \int d\underline{\Omega} \int dE' \Sigma_s(E' \rightarrow E, \underline{\Omega}' \rightarrow \underline{\Omega}, \underline{r}) \; \phi(\underline{r},E',\underline{\Omega}') + S(\underline{r},E,\underline{\Omega}) \qquad (1.1)$$

Here the basic variable is the angular neutron flux at position $\underline{r}$, energy E and in a direction $\underline{\Omega}$, indicated by $\phi(\underline{r},E,\underline{\Omega})$

$\Sigma_s(E,\underline{r})$ is the neutron scattering cross-section at energy E and position $\underline{r}$,

$\Sigma_a(E,\underline{r})$ is the corresponding neutron absorption cross-section.

$$\Sigma_s(E' \rightarrow E; \underline{\Omega}' \rightarrow \underline{\Omega}; \underline{r}) = \Sigma_s(E', \underline{r}) P(E' \rightarrow E, \underline{\Omega}' \rightarrow \underline{\Omega}, \underline{r})$$

where $\Sigma_s(E', \underline{r})$ is the neutron scattering cross-section at energy E' and position $\underline{r}$, and $P(E' \rightarrow E, \underline{\Omega}' \rightarrow \underline{\Omega}, \underline{r})$ is the probability that a scattering collision at position $\underline{r}$ by a neutron with energy E' and direction $\underline{\Omega}'$ will cause the neutron to change its energy to E and its direction to $\underline{\Omega}$.

1

$S(\underline{r},E,\underline{\Omega})$ represents the source of neutrons (through fission or fixed sources).

The equation is most easily thought of as a neutron balance equation with the lefthand side dealing with processes tending to remove neutrons from the element at $(\underline{r},E,\underline{\Omega})$ and the righthand side with processes tending to restore them. Many books(1,2) discuss approximate methods of solution to this equation. Most concentrate their attention on the spherical harmonic method of solution, in which ϕ, the flux is expanded as a series of spherical harmonic functions so that we write:

$$\phi(\underline{r},E,\underline{\Omega}) = \phi(\underline{r},E) \sum_{\ell=0}^{\infty} \sum_{m=-\ell}^{\ell} g_{\ell m} Y_{\ell m}(\underline{\Omega}) \qquad (1.2)$$

and curtail the expansion after a finite number of terms in the series, which amounts to making a specific assumption about the angular form of the flux.

The appeal of such an approximation is two-fold. First, if we make the particularly simple assumption that we can terminate the series after the first term, and similar approximations(1) about the scattering kernel, the equation reduces to the so-called diffusion approximation:

$$- D\nabla^2 \phi(\underline{r},E) + \Sigma(E,\underline{r}) \phi(\underline{r},E)$$

$$= \int_{0}^{\infty} dE' \Sigma(E' \to E) \ \phi(\underline{r},E') + S(\underline{r},E) \qquad (1.3)$$

which was originally used for most theoretical reactor analysis and is still used extensively where the spatial variation of the neutron flux is slow (in whole reactor calculations for example). Second, because this first approximation is so well known and understood, it is possible to see the higher harmonic terms as corrections to diffusion theory and to gain a physical feel for the meaning of the higher-order terms.

However, the space devoted to such methods should not blind us to the fact that reactor design is, and largely has been, dominated by the multi-group Carlson(3,4,5) S_n method

of solution for the last 20 years or so. The multigroup
formulation is a standard procedure, starting from Equation
(1.1). The energy range is divided into a series of non-
overlapping segments $\{E_1, E_2, E_3 \text{----} E_{n+1}\}$ and Equation 1.1
is subjected to the operator

$$\frac{1}{E_{r+1} - E_r} \int_{E_r}^{E_{r+1}} dE$$

Under these conditions, the equation reduces to an equation
for each group:

$$\left[\underline{\Omega} \cdot \nabla + \Sigma_{tg}(\underline{r}) \right] \quad \phi_g(\underline{r}, \underline{\Omega}) = Q(\underline{r}, \underline{\Omega}) \tag{1.4}$$

where the source term $Q(\underline{r}, \underline{\Omega})$ is given by

$$Q(\underline{r}, \underline{\Omega}) = \int d\underline{\Omega}' \left[\sum_{g'} \Sigma_{sg'g}(\underline{r}, \underline{\Omega}' \to \underline{\Omega}) \phi_{g'}(\underline{r}, \underline{\Omega}) \right] + S_g(\underline{r}, \underline{\Omega}) \tag{1.5}$$

$\phi_g(\underline{r}, \underline{\Omega})$ is the angular flux in the gth energy group,

Σ_{tg} is some average of the total cross-section in group
g as a function of energy,

$\Sigma_{sg'g}$ is the scattering cross-section from group g' to
g, and

$S_g(\underline{r}, \underline{\Omega})$ represents the fixed source into the gth energy
group.

Equation (1.4) is a particularly convenient form of the
equation. It is basically of one group form, with the righthand side representing neutron flow into the energy group g
from other groups and fixed external sources. It also lends

itself conveniently to iterative methods,\ since the fluxes ϕ_g, when calculated, can be used to update calculations of Q. This process eventually converges to the group fluxes

$$\phi_1, \phi_2, \text{ --- } \phi_g.$$

This still leaves the problem of the angular integration. In the Carlson(3) method, the integral in the transport equation is approximated by the device of splitting the angular interval up into equal subintervals and assuming a linear variation of the flux in each subinterval. Suprisingly little development of the method has occurred (see Askew and Brissenden(5), for example) and the approximation still is being used essentially in its original form.

Besides this extensive work on the integro-differential Boltzmann equation, the other substantial progress has been through the development of collision probability methods, usually working from the integral transport equation

$$\phi(\underline{r}, E, \underline{\Omega}) = \int dR' \exp \left\{-\int_0^{R'} \Sigma(E, \underline{r} - R''\underline{\Omega}) dR''\right\} Q(\underline{r} - R'\underline{\Omega}, E, \underline{\Omega})$$

which effectively calculates the flux at $(\underline{r}, E, \underline{\Omega})$ by summing the contributions to it from neutrons traveling from all points in space. Collision probability methods are particularly valuable in dealing with complex cluster geometries where the detailed representation of the fuel is of importance(6).

The method described here is new, and yet in some sense seeks to combine the advantages of both types of solution outlined here. Essentially, what is done is to take a collision probability (i.e., a neutron tracking) approach to the integro-differential form of the Maxwell-Boltzmann equation instead of the integral form. An outline of the method is given in Section II, and it is illustrated by applying it to two fairly simple cases (the slab and the sphere). For the important practical case of cylindrical geometry, a computer code CACTUS(7) has been written. This is described in Section III including details of the problems (mostly geometric) met during implementation. Section IV compares results obtained by the characteristic methods with those of standard methods applied to 'benchmark' problems. Finally Section V estimates the value of the method of characteristics and suggests future developments and applications.

II. THEORY OF CHARACTERISTICS METHODS AND TWO EXAMPLES

A. Notation and General Equations

We shall find the mean angular flux within all geometric regions of interest for each of a discrete set of angles Ω_1, Ω_2, ... Ω_j, ... Ω_J. In order to avoid excessive indices, we suppress any reference to which energy group we are considering. The mean cross-section for all processes within a particular region, labelled by i, will be supposed known and equal to Σ_i. The geometry and nuclear properties will be modeled by defining each geometric region, taking Σ_i to be constant within region i. Scattering and fission cross-sections, required to calculate the source function, also will be taken to be constant within a region.

Introducing a coordinate system (x,y,z), so that $x = \underline{r} \, \underline{\Omega}_j$, the Maxwell-Boltzmann equation (1.4) becomes:

$$\Sigma_i \; \phi_j(x,y,z) = Q_j(x,y,z) - \frac{d\phi_j(x,y,z)}{dx} \qquad (2.1)$$

where we have introduced $\phi_j(x,y,z) = \phi(\underline{r},\underline{\Omega}_j)$ and

$$Q_j(x,y,z) = Q(\underline{r},\underline{\Omega}_j).$$

This total differential equation is the fundamental equation of the characteristics method of solving the transport equation. The straight lines defined by y = constant, z = constant are the characteristics for Equation (1.4).

To obtain $\bar{\phi}_{ij}$, the mean flux for this angle over region i, we integrate over its volume V_i.

$$V_i \, \bar{\phi}_{ij} = \int_{V_i} \phi_j(x,y,z) \; dxdydz \qquad (2.2)$$

This integration is carried out by considering the region to be made up of a series of cylinders, with axes parallel to $\underline{\Omega}_j$ (it may be that the region is reentrant, in which case some cylinders will be made up of several disconnected sections). Within each cylinder, we find the mean angular flux along some representative line, parallel to the axis, summing over all cylinders, with weights proportional to volume, to obtain $\bar{\phi}_{ij}$. The way we choose the representative line will depend upon the particular problem being studied. In the general three-dimensional problem, it would be chosen to pass through the centroid of the cylinder.

Suppose the line through the k'th cylinder, cross-sectional area, A_{jk}, passes through the point $(0, y_k, z_k)$ and has length L_{ijk} through region i. $\bar{\phi}_{ijk}$, the mean value of ϕ_j along this line, is given by:

$$L_{ijk}\, \bar{\phi}_{ijk} = \int_{L_{ijk}} \phi_j(x, y_k, z_k)\ dx \qquad (2.3)$$

This gives, on substituting for ϕ_j, using Equation (2.1):

$$\Sigma_i L_{ijk}\, \bar{\phi}_{ijk} = \int_{L_{ijk}} Q_j(x, y_k, z_k)\ dx -$$

$$- \int_{L_{ijk}} \left\{ \frac{d\phi}{dx}_{\cdot j}(x, y_k, z_k) \right\}\ dx,$$

or $\quad \Sigma_i\, L_{ijk}\, \bar{\phi}_{ijk} = \int_{L_{ijk}} Q_j(x, y_k, z_k)\ dx - \phi'_{ijk}, \qquad (2.4)$

where

$$\phi'_{ijk} = \phi_j(x_1, y_k, z_k) - \phi_j(x_2, y_k, z_k),\ \text{with}\ (x_1, y_k, z_k),$$

and (x_2, y_k, z_k) being the points where the line enters and leaves the region, respectively.

Summing, with weights proportional to $L_{ijk} A_{jk}$, as mentioned above;

$$\bar{\phi}_{ij} \quad \sum_k A_{jk} \int_{L_{ijk}} Q_j(x, y_k, z_k)\, dx \Big/ \sum_k A_{jk} L_{ijk} + \qquad (2.5)$$

$$+ \sum_k A_{jk} \phi'_{ijk} \Big/ \sum_k A_{jk} L_{ijk}$$

The first term on the righthand side of Equation (2.5) is
an approximation to $\bar{Q}_{ij}$, the mean source term for angle j
in region i, so we may write (2.5) as:

$$\bar{\phi}_{ij} \simeq \bar{Q}_{ij} + \sum_k A_{jk} \phi'_{ijk} / \overset{+}{\sum_k} A_{jk} L_{ijk}, \qquad (2.6)$$

an equation that holds exactly in the limit of the cylinders having zero cross-sectional area. In fact, because of
the approximations required in calculating the ϕ'_{ijk}, it is
better to have small regions with few lines through each,
and in this case, $\sum_k A_{jk}L_{ijk}$ is not a very good approximation to the volume of region i, V_i. There are good reasons
for getting the tracked area correct for each angle. In
particular, the number of neutrons captured in a region is
proportional to its volume, as are the number of fissions,
so calculation of integral properties like k_{eff} should use
correct volumes. We can arrange to get correct volumes for
each region for angle j by scaling the lengths L_{ijk}, as long
as at least one line passes through each region for this
angle. The scaled lengths L'_{ijk} are given by:

$$L'_{ijk} = V_i L_{ijk} / \sum_k A_{jk}L_{ijk} \qquad (2.7)$$

If we then calculate the ϕ'_{ijk} using L'_{ijk} instead of L_{ijk},
calling the ϕ'_{ijk} obtained Δ_{ijk}, we have:

$$\bar{\phi}_{ij} = \bar{Q}_{ij} + \sum_k A_{jk}\Delta_{ijk}/V_i \qquad (2.8)$$

Equation (2.8) is used iteratively; a starting set of mean
fluxes are used to derive the $\bar{Q}_{ij}$, using Equation (1.5) Note
that this equation holds, by linearity, not just at a point,
but for mean values throughout a region with constant nuclear
properties), the Δ_{ijk} found, and Equation (2.8) used, to give
a better approximation to the $\bar{\phi}_{ij}$.

Within each region we find the Δ_{ijk}'s by integrating
Equation (2.1). We will have, generally, the value of ϕ_j
(x_1,y_k,z_k), from boundary conditions for the first region or

as the previous $\phi_j(x_2,y_k,z_k)$ for successive regions, and this, together with some suitable model for ϕ_j, Q_j within the region allows solution.

One particularly useful approximation is $Q_j(x,y_k,z_k) = \bar{Q}_{ij}$ within region i. This flat source approximation gives:

$$\phi_j(x_2,y_k,z_k) = \phi_j(x_1,y_k,z_k)\exp(-\Sigma_i L'_{ijk})$$

$$+ \bar{Q}_{ij} \{1 - \exp(-\Sigma_i L'_{ijk})\} /\Sigma_i \qquad (2.9)$$

To date, this is the approximation used in studies at Bristol, but Askew(8) has suggested that, for larger regions, the approximation of constant neutron current can be used to advantage.

B.   Slab Geometry

Suppose we stack a layer of finite plates, each of constant thickness and material properties. If the dimensions of each plate are very large, compared with the neutron mean free path, then we may regard each as an infinite plate. This so-called "slab geometry" (Figure 3) will have neutron fluxes as a function of x and θ only, where the X direction is chosen perpendicular to the surfaces and $\cos\theta = \vec{OX}.\vec{\Omega}$. We shall use this problem as an illustration of the method of characteristics. We shall find the $\bar{\phi}_{ij}$, as in (2.1), summing these with appropriate weights w_j, to give approximate mean scalar fluxes $\bar{\phi}_i$ in each region i.

As the flux depends on x and θ only, it is sufficient to track one line through the system for each angle j. We may suppose we start tracking at x = 0 (if $\theta > \pi/2$, start similarly at the righthand boundary). We then take some value for $\phi_j(0)$ and track through, as in Section II A, giving:

$$\bar{\phi}_{ij} = \bar{Q}_{ij} + \phi'_{ij} (\delta x_i \sec\theta_j), \qquad (2.10)$$

where $\delta x_i = x_{i+1} - x_i$ is the thickness of the ith slab (see Figure 1). This equation is now exact, and the only approximation is in finding the ϕ'_{ij}. If we use the flat source approximation, we obtain:

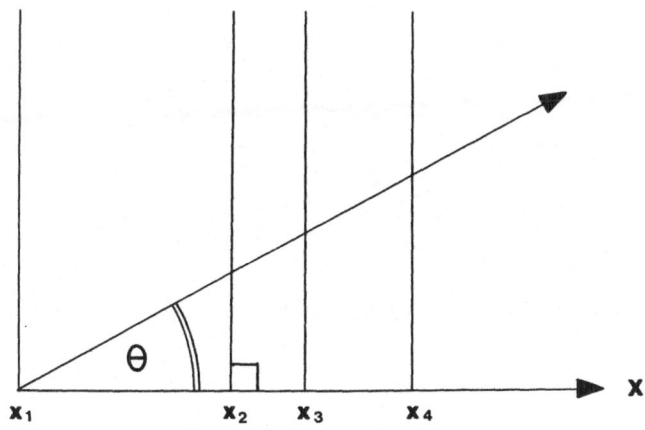

Figure 1.   Slab Geometry

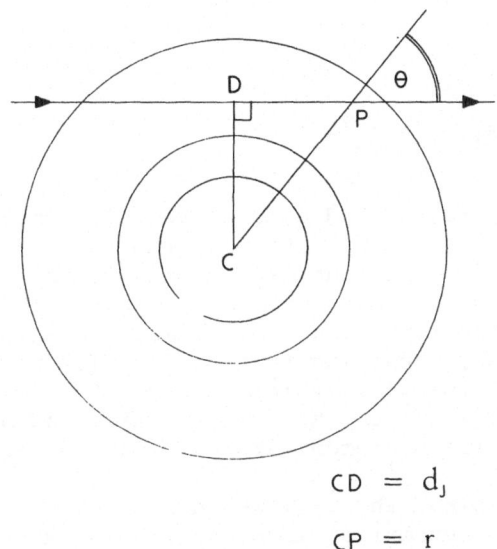

$$CD = d_j$$

$$CP = r$$

Figure 2.   Spherical Geometry (r, θ)

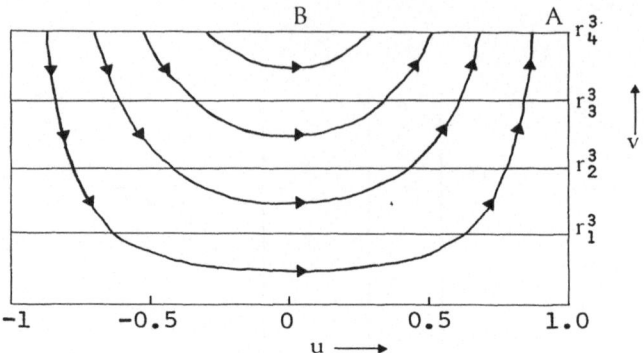

Figure 3.   Spherical Geometry (u,v)

$$\phi_j(x_{i+1}) = \phi_j(x_i) \exp(-\Sigma_i \delta x_i \sec\theta_j) \tag{2.11}$$

$$+ Q_{ij} \{1 - \exp(-\Sigma_i \delta x_i \sec\theta_j)\}/\Sigma_i$$

with $\phi'_{ij} = \phi_j(x_{i+1}) - \phi_j(x_i)$

The slabs are treated successively for this angle until the line leaves the last plate. In the case of a finite stack of plates with no incident neutrons, where we have $\phi_j(0) = 0$, we then go on to the next angle. In the case of a periodic system with length a, we may use $\phi_j(a)$ as $\phi_j(0)$ for another scan, repeating this until $\phi_j(a)$ converges. It is probably computationally more efficient, however, to update the source functions at the same time and so the strategy will be little different from the finite case.

Having obtained the $\bar{\phi}_{ij}$, we then combine them to give $\bar{\phi}_i$. Unlike many numerical methods of solving the Maxwell Boltzman equation (1.4), we are totally at liberty to choose any set of angles θ_j, with weights determined from an appropriate model and angular dependence. This freedom may be

very important in studying problems where the angular dependence is expected to be fairly singular. Thus, in studying
penetration problems, we are able to choose a greater density
of angles in the X direction, where the flux will be much
greater than for $\theta \geq \pi/2$. In the Meneghetti effect(9), we
might choose many angles near $\theta = \pi/2$, where chaneling occurs
in some slabs. There is the possibility that an approximate
solution might be used to choose a better set of tracking
angles and repeated iteratively, but this has not been investigated as yet.

Suppose we have chosen a set of angles. How do we determine the weights? One obvious choice is to take ϕ to be a
linear function in θ; then we get for the scalar flux:

$$\phi(x) = \int_0^\pi \phi(x, \theta) \, \sin\theta \, d\theta$$

$$= \int_0^\pi \sin\theta \{\phi_j(x) + (\phi_{j+1}(x) - \phi_j(x))(\theta - \theta_j)/(\theta_{j+1} - \theta_j)\} d\theta \tag{2.12}$$

Integrating and collecting terms, we obtain

$$\phi(x) = \sum_j w_j \phi_j(x) \tag{2.13}$$

so the w_j are the required weights.

C.   Spherical Geometry

In the case of a system that has rotational symmetry
about a point, the neutron flux will depend only on r (the
distance from the center) and the angle between the direction
of the angular flux and the radius vector. The obvious geometric model to use is one of a number of shells (like an
onion), with the ith region being the points whose radius r
satisfies $r_{i-1} < r < r_i$, with $0 = r_0 < r_1 < r_2 < r_N = R$.
We shall consider the characteristics method of finding ϕ_{ij},
angular fluxes or $\bar{\phi}_i$, scalar fluxes for each of the regions.

If we track along straight lines as shown in Figure 2, for the jth line, $r \sin \theta = d_j$. This line will pass through some or all of the shells, with a different mean value of within each shell. One way of visualizing this is to plot these lines in u,v coordinates, where

$$u = \cos\theta, \qquad v = r^3.$$

The equation of one line is then

$$v = d_j^3 (1 - u^2)^{-3/2}$$

A system with four shells and four lines arranged so that the jth line passes through $4 - j$ shells is shown in Figure 3.

Near A, we have a high density of tracking lines in phase space. This point corresponds to outward directed lines at the surface, and we would hope to have better angular modeling here than at B, corresponding to tangents to the system. Only one line passes through the innermost region, but hopefully we have chosen this region small enough for the flux to be reasonably isotropic.

Let us suppose that the angular fluxes we obtain along the line j are to be used as representative for the regions between the cylinders $r \sin\theta = r_{i-1}, r_i$. We first must decide where to put d_j in the interval (r_{i-1}, r_i), then track to find the ϕ_{ij} as previously, and finally combine with appropriate weightings to give ϕ_i.

One approach which gives d_j is to arrange that they are defined successively, so that the tracked volume of each shell is correct. This has been suggested by Askew (10). To illustrate this, return to the four-shell situation again. The innermost 'tube' is obtained as the intersection of a cylinder radius r, with the system (analogous to pushing an apple corer through an onion, Figure 4). The cross-sectional area of this cylinder is r_1, so if we take $d_1 = \sqrt{5} r_1 / 3$, the length of this line through region 1 will be $4r_1/3$, giving the correct volume for this region. In addition, this will define contributions to each of the other regions. To deal with region 2, pass a larger cylinder, radius r_2 through the system. We now have a tube for region 2, together with the two caps already obtained from cylinder 1. The volume of the caps has been fixed already by choosing d_1, so we choose d_2 then the total volume for the three is

$$\frac{4}{3} \pi (r_2{}^3 - r_1{}^3).$$

The dissection of region 2 is shown in Figure 5. This idea can be continued for the remaining regions.

If we take zero incident flux, then we start tracking the jth line at the boundary, with $\phi_j = 0$. L_{ij}, the length of the jth line through the ith region, is given by:

$$L_{ij} = \sqrt{r_i{}^2 - d_j{}^2} - \sqrt{r_{i-}{}^2 - d_j{}^2} \quad \text{for } j < i \qquad (2.14)$$

$$L_{ij} = 2\sqrt{r_i{}^2 - d_j{}^2} \qquad \text{for } j = i$$

$\bar{\phi}_{ij}$, given by

$$\bar{\phi}_{ij} = \{ \int_{L_{ij}} Q_j(x)\, dx + \Delta_{ij} \} / \Sigma_i L_{ij} \qquad (2.15)$$

gives the mean flux along the jth line through region i. This line passes through the shell at an angle θ to the radius vector, ranging from $\sin^{-1}(d_j/r_i)$ to $\sin^{-1}(d_j/r_{i-1})$ (or $\sin^{-1}(d_j/r_i)$ to $\pi/2$ if i = j). By the mean value theorem, $\bar{\phi}_{ij}$ is the mean angular flux at some point in the region. By decreasing the size of shells, we are, of course, able to reduce this indeterminacy.

The $\bar{\phi}_{ij}$'s obtained may be used to generate $\bar{Q}_{ij}$'s, giving an iterative method of solving for the $\bar{\phi}_{ij}$'s. In the case of isotropic scattering, there is no need to find the $\bar{\phi}_{ij}$'s to obtain the source function, so it is possible to find $\bar{\phi}_i$ directly, with a corresponding computational saving. To decide the weights we should associate with each line, note that for fixed r, the neutron flux through this radius within a given tube is proportional to the cross-sectional area of the tube (assuming isotropic flux). We therefore associate weights proportional to the cross-sectional area of

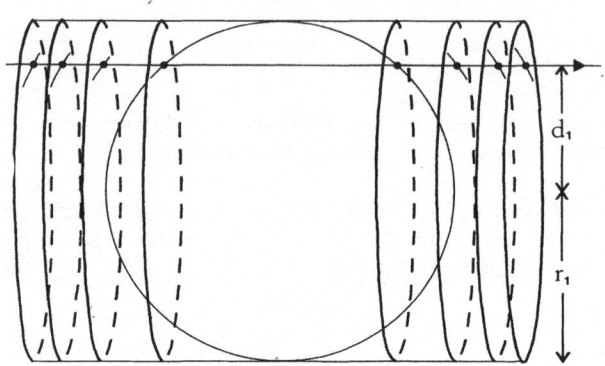

Figure 4.   Innermost Cylinder in Spherical Geometry

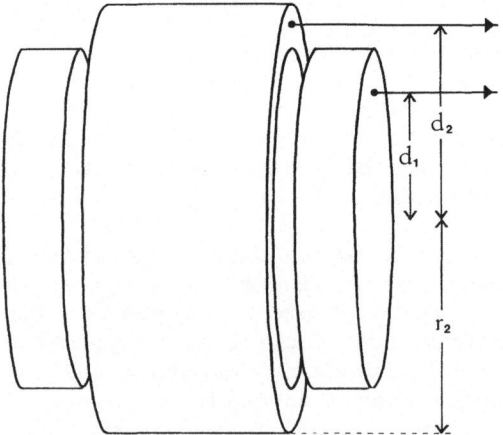

Figure 5.   Dissection of Second Shell in Spherical Geometry

each tube, giving:

$$\bar{\phi}_i = \bar{Q}_i + \sum_j A_j \Delta_{ij} / V_i,$$ (2.16)

where

$$A_j = \pi(r_j^2 - r_{j-1}^2)$$ (2.17)

Askew(10) has given a one-group spherical problem with radii 2,4,6 cm; Σ_s = 0.05, 0, 0.1; Σ_t = 0.05, 0, 0.5 and fixed fission source 1, 0, 0. The angular flux at the surface is shown in Figure 6. This method deals with the rapid change in the flux in the region of tangent lines to the fission source.

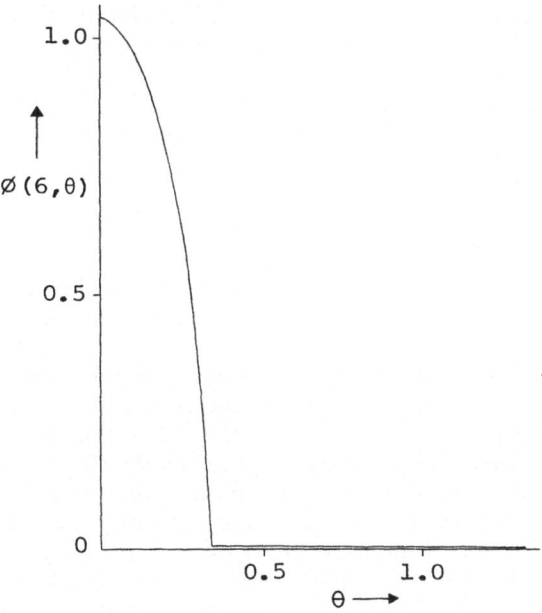

Figure 6.   Angular Flux at Surface of Spherical Problem

III. CYLINDRICAL GEOMETRIES: THE CACTUS CODE

A. Introduction

For a given real reactor, it is difficult to treat both
the small scale variation of neutron flux from region to
region and the overall variation due to non-uniform loading
and increased leakage probability near the boundary. Fortunately, because of engineering considerations, the geometry often is of a particular type, which makes it possible
to use a synthesis method. It is clearly advantageous to
design reactors to be made up of a minimum number of types
of element so there is often a basic element of structure
replicated throughout the reactor. In a particular horizontal plane, therefore, we might expect a superimposition
of the fine variation within a unit cell onto an overall
flux shape to be a useful approximation. Further, again for
these regions, and more particularly, because of loading and
fluid flow considerations, the geometry is to a large degree
the same wherever a horizontal "cut" is taken through the
system (because of the axial decrease in flux toward the
boundary, aging effects will vary with height in the reactor
but this will be a smooth function of height). This section
will consider how the method of characteristics may be used
to solve for neutron flux levels within a periodic (infinite)
lattice on cylindrical geometry and some comments on experience in devising a computer code to carry this out.

B. Characteristics Used

As the system is infinite, we have characteristics that
will be of infinite length. This problem may be resolved
in one of two ways. The first is to suppose that for each
angle the neutron flux on the boundary of the unit cell is
given by a function involving parameters that can be updated as calculation proceeds. This idea has been used in
transport calculations before (e.g., by Mordant(12)). Alternately, the tracking scheme originally suggested by Briss-
enden(11), or the rational tangents scheme, for collision
probability calculations may be used.

For a square unit cell in the (x,y) plane, the Brissenden scheme uses lines that satisfy

$$\frac{dy}{dx} = \frac{p}{q},$$ with p and q relatively prime integers. Suppose

we take a unit cell and $p = 2$, $q = 3$, for example; any
characteristics then will be periodic, returning to the
same point after crossing the boundary of the unit cell
five times. In Figure 7, such a characteristic is shown,
together with the reduction to the unit cell. In cases
where the unit cell has symmetry, we can use a smaller cell
to advantage, but the characteristics then are a little more
complicated. For a single line of symmetry in the x-y plane,
taken to be parallel to the x-axis, we have the situation as
in Figure 8. Figure 9 is that obtained with eightfold symmetry (reflection in diagonal and lines parallel to the x
and y axes). This eightfold symmetry is a feature of the
benchmark problems discussed in the next section. A line
starting at 1, with θ_j in the range $(0, \frac{\pi}{4})$ (where θ is the
angle between the characteristic and the x-axis, hereafter
called the polar angle) will be reflected within the $\frac{1}{8}$
unit cell to give angles in the range $(0, 2\pi)$. The variation
of flux with θ in the unit cell may be modeled by choosing
starting angles in the range $(0, \frac{\pi}{4})$. Clearly, there is some
restriction on the polar tracking angle, being solutions to θ_j
$= \tan^{-1}(\frac{p}{q})$, with p,q integers, but for large p and q, this
gives a large number of possible tracking angles. The larger the value of $p + q$, the finer is the mesh of lines generated, with corresponding improvement in the approximation
to the geometry this gives. Figure 10 shows a square unit
cell enclosing a single circular region, with the approx-
imation when $\frac{p}{q} = \frac{3}{7}, \frac{9}{19}$. The distance between nearest parallel
lines will be referred to as the spatial mesh separation.

C.   Angular Quadrature

     The most obvious approach is to divide the surface of
the sphere up into equal areas. An example of this is in
Mordant(12). Each angle then will have an equal weight in
forming the scalar fluxes. An alternative is to take a
Cartesian product of two sets of angles $\theta_1, \theta_2 \ldots \theta_j \ldots \theta_{N_j}$
and
     $\lambda_1, \lambda_2 \ldots \lambda_k \ldots \lambda_{N_k}$,
where θ_j is the jth polar angle and λ_k the kth azimuthal
angle (angle between the characteristic and the x-y plane).
This has the advantages that tracking need only be carried
out for each of the N_j polar angles and the geometrical
approximation due to finite spatial mesh separation is the
same for all azimuthal angles associated with a particular
polar angle. Used in conjunction with a method of saving

tracked lengths and possibly exp $\{-\Sigma_{ig} L_{ijkm}\}$, where Σ_{ig} is
the total cross-section for region i in energy group g, and
L_{ijkm} the length of the mth line through region i for angle
(θ_j, λ_k), this may result in great economy of calculation.
In principle, we should choose a different angular quadrature
for each energy group, those where the fluxes are more nearly
isotropic requiring less angular resolution. However, to date,
this has not been investigated, a single angular quadrature
scheme being used for all groups in the CACTUS code. In this
case, we need only track N_j lines, other tracked lengths being
found by

$$L_{ijkm} = L_{ijom} \sec\lambda_k, \tag{3.1}$$

where L_{ijom} is the corresponding length of the projection
in the x-y plane. As the geometric approximation is the same
for each k, we may use a linear model of flux with azimuthal
angle, giving weights associated with each k (as in the slab
case). We also may use the fact that

$$\phi(r, \theta, \frac{\pi}{2}) = \frac{Q(r)}{\Sigma_i} \tag{3.2}$$

within region i (group indices being suppressed) for isotropic
scattering.

It has been suggested by Askew(8) that within a general
Cartesian structure, we might choose a subset of the N_j
angles for each value of k. This has the advantage of allowing more nearly equal solid angles with each tracked angle,
but the disadvantage that we no longer can use linear integration over λ.

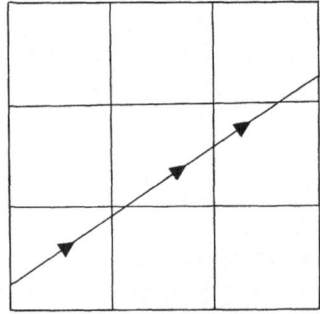 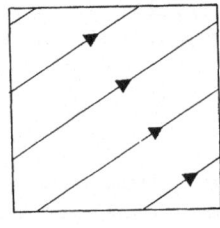

Figure 7.   Characteristic Ray Solutions of the
            Transport Equation

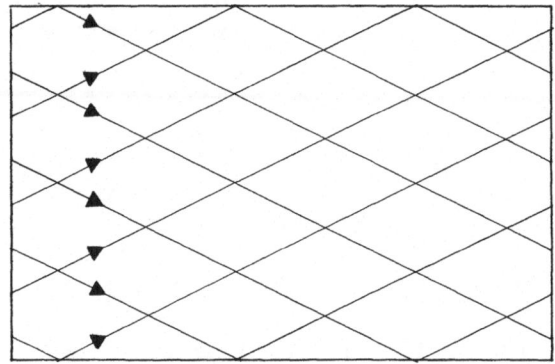

Figure 8.   Characteristics in Rectangular Cell
            with Two-fold Symmetry

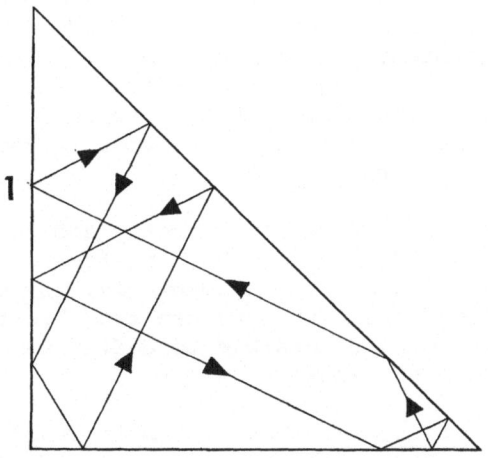

Figure 9.   Characteristics in Rectangular Cell
            with Eight-fold Symmetry

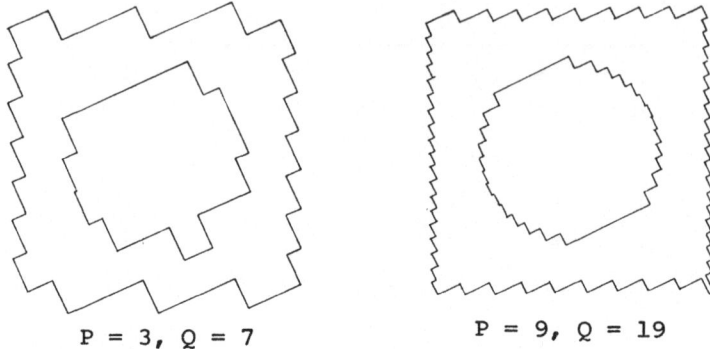

P = 3, Q = 7   P = 9, Q = 19

Figure 10.   Two Approximations to a Simple Cell

D.   Overall Strategy for a Particular Code - CACTUS

Within one group, the geometry has been reduced to that
of the slab case for each angle (this reduction of the spatial
scan required to one dimension leads to the possibility that,
for sufficiently large systems, the method of characteristics
may be more efficient than other methods). Some of the slabs
are identified, according to the region in the x-y plane from
which them came. Therefore, we proceed as for the slab case,
except that there is an additional outer iteration corresponding to between-group scatter.

The interface between the geometry and the flux calculation is just the tracked lengths, together with their lengths
and weights for each angle. Accordingly, any geometric data
structure may be defined that we can track in to give these
lengths. It has proved possible to define a very general
geometry as outlined below.

Geometric Structure. The system is described by defining "elements". Each element is defined by a finite number
of "nodes" (usual sense) joined by arcs or straight lines.
(the number of sides has been arbitrarily limited to up to
8). In addition, non-intersecting annular systems may be
defined within the element. If it seems likely that a large
neutron flux tilt occurs across an element, it could, of course
be subdivided.

The benchmark calculation(*13*) referred to in the next
section was studied using the two geometric meshes shown in
Figure 11.   To track in this geometry, we have to track a
given line from when it enters an element to when it leaves
it, calculating line segment lengths.   In addition, we must
be able to deal with boundaries (either by translation or
reflection, depending upon symmetry).   It is easy to do this,
the problems coming with lines that are very close to nodes
or tangents to arcs.   It was found that sometimes numerical
roundoff could cause two intersections in such cases to be
reversed in order, causing the line to get "lost".   This was
dealt with by moving the line up or down a small amount when
near a node or tangent, in order to move away from it.   This
will, of course, not affect solutions, as the length thus
ignored will be very small.

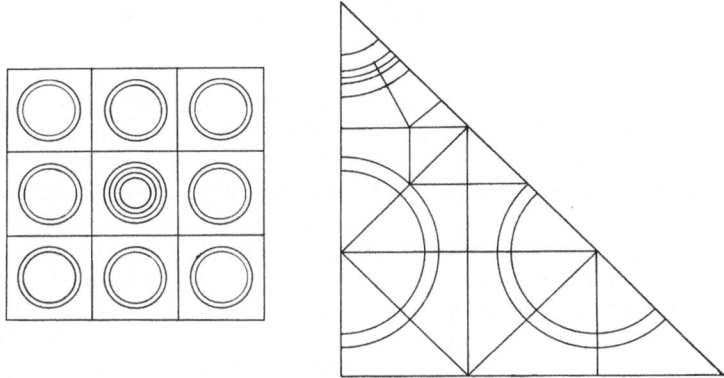

Figure 11.   Geometry Used in Benchmark Problem 1

As previously mentioned, the tracked areas in the x-y
plane were not, in general, correct.   This is undesirable
because the total number of removals depends upon the tracked
areas, whereas the fission source is calculated from line
areas.   The lengths, therefore, are scaled so that the tracked
areas are correct for each angle (the areas being automatic-
ally calculated on defining the geometry).

E.   Convergence

Each outer iteration of the fission source terms
for each region are recalculated from the latest scalar
fluxes, scaling so that the total number of fissions is
constant and equal to 1. For each group (working downward
from the fast groups) source functions were calculated and
Δ's calculated as previously. The angular flux "in" is
taken to be the last one obtained "out" for this angle. This
difference, summed with the appropriate weights over all
angles, gives a measure of the overall neutron balance in
this group. The sum of squares of flux changes by region,
weighted by volume of region, is also found. These two
parameters are compared with an estimate of the fission
source errors to decide whether to go on to the next group
or carry out another iteration on this group. Two constants
may be defined to alter this ratio of inner-outer convergence.
In practice, this works well for groups where there is little
upscatter. Some investigation has been made of allowing a
freer iteration strategy and it seems likely that any future
work could benefit from this.

There is another feature, however, that is more important in determining solution time. For a given geometric
mesh, the time for one inner iteration is proportional to the
total number of angles used and inversely proportional to the
tracking mesh used. It has proved advantages to gradually
step up each of these, using the previous fluxes as starting
values. In addition, before calculating any scalar fluxes,
the ingoing angular fluxes are calculated by tracking through
the system, without updating scalar fluxes. This, besides
giving a great time saving, has enabled an idea to be formed
of the effect the mesh used has on the result. This distinguishes the method of characteristics from more conventional
collision probability methods, where the major time is spent
effectively inverting the collision probability matrix; the
time taken to do this does not depend upon the tracking
finesse used, merely on the number of regions. The characteristics method gets transfers between regions i,i' much more
accurately when they are "near" than when they are more remote, whereas, collision probability methods treat all pairs
of regions the same. In this sense, the characteristics
method is much more like the Monte Carlo method.

There is another sense in which it may be compared to Monte Carlo methods. Perturbation problems have been studied using correlated Monte Carlo methods. If the characteristics method is to be used, some thought will indicate that the same angular and tracking mesh should be used, as the non-deterministic change in shape and tracked volumes would mask smaller differences.

IV. COMPARISON WITH OTHER METHODS

A. Introduction

Exhaustive testing of a new method such as that outlined here is always a lengthy process. It is difficult to test such a method against experiment since present experimental methods do not reveal the detail predicted by the model, and hence, form only a crude test. It is clearly a necessary condition that the method should predict experimental results correctly, but it is not sufficient. A similar situation exists when comparing the predictions of the model in cases where an exact solution is known. For example, the Milne problem(1) is only exactly soluble in a small number of cases, and usually for situations that bear little relation to the real world; e.g., all neutrons at the same speed, separate isotropic kernel, etc. Again, it is necessary that a new method reproduces these results, but hardly sufficient.

Fortunately, at just the time the problem of testing this particular method occurred, a readymade solution occurred. It was agreed in 1974 at the Nuclear Energy Agency Committee to set up a series of "benchmark" calculations, covering typical-but demanding-situations in BWR's. Detailed geometry and cross-sections were provided so that the various participating establishments could compare methods, and avoid the usual discrepancies caused by the use of a different geometrical description or different cross-sections.

The results of this exercise are discussed in detail by Halsall(13). The four standard benchmarks were:

1. a poison pin supercell

2. a "mini-BWR" problem, with four pin cells containing a range of enrichments and a simulated cruciform control rod

3.   a full-size BWR lattice, with differing
     fuel enrichments and poisoned fuel pins

4.   a similar lattice to that described in
     (3), but containing also, an empty pin
     position and a cruciform rod

Detailed calculations have been carried out for the
first two benchmarks, and results are given below. One problem in making comparisons is that the calculations are not
"benchmarks" in the sense that a correct answer exists and
is known, but Halsall concludes that the two methods submitted originally, which had the most detailed solution
methods are in good agreement, and are hence considered to
provide the "true" solution. These two detailed calculations
are provided by A. E. E. Winfrith (AEEW) and the Japanese
Atomic Energy Research Institute (JAERI) and are referred
to by these initials in the comparisons that follow.

B.   Poison Pin Supercell

The geometry for this problem has been discussed already (Figure 11).

For each problem, the dimensions, material layout and
six group macroscopic cross-sections for each material were
provided. The output requested was the infinite lattice
multiplication, absorptions by material and maps of fission
rate, and of groups 4 and 6 fluxes. Input data are not included here, since it is easily available from Halsall.

Clearly, a great deal of symmetry is in the problem
(eightfold to be exact), and essentially, there are only
three different types of pins, namely:

1.   the central poison pins

2.   fuel rods in corner cells

3.   fuel rods in the middle of sides

This simplifies the absorption and flux output considerably, since it need be given only for each type cell, of which
there are only two distinct types in this problem, although
it can be divided into fuel, can and coolant results within
each cell. Absorptions given by the present (CHARACTERISTIC)

method are compared with best AEEW and JAERI results in
Table I.

TABLE I

		AEEW	CHAR.	JAERI
OUTER	FUEL	.7269	.7254	.7241
PIN CELL	CAN	.0106	.0105	.0105
	COOLANT	.0226	.0223	.0223
INNER	FUEL	.2364	.2381	.2395
PIN CELL	CAN	.0012	.0012	.0012
	COOLANT	.0023	.0024	.0024

This shouws good agreement with what are considered to
be the best results, lying between, in fact. Detailed results also are available for flux maps for Groups 4 and 6,
but these are not included in the interests of not submerging
the paper in detail.

One interesting point pursued in some detail for this
fairly simple calculation has been the sensitivity of the
output to variations in the details of the models used in
the characteristics method. For example, the following variables are at the disposal of the person using the code:

1. separation of the tracking lines

2. number of polar angles

3. number of aximuthal angles

Confining our comparisons of the effect of these only
to absorptions in the fuel region, we get the variations
shown in Table II.

TABLE II

SEP. OF Tr. LINES	POLAR ANGLES	AZIMUTHAL ANGLES	OUTER FUEL ABS.	INNER FUEL
0.1	1	9	.7256	.2387
0.1	3	9	.7241	.2395
0.1	5	9	.7254	.2381
0.15	3	6	.7234	.2402
0.07	3	9	.7254	.2383

Clearly, within the accuracy we are seeking to achieve here, the results are relatively insensitive to details of the models chosen. This is rather less true of the fluxes, for example, where a coarse tracking mesh can lead to only one line (say) passing through a particular region, and this poor sampling can lead to inaccuracies. This is a defect of collision probability codes in general, of course. The problem of variation of azimuthal angles has not been considered in detail, but of course this is common to all transport codes, and is not dealt with differently in the characteristics method.

C.   BWR with Cruciform Rod

The geometry for benchmark problem No. 2 is shown in Figure 12. This problem, and particularly the presence of the cruciform control rod, is a particularly exhaustive test for any code. Here, there are essentially three types of pin cell:

1.   closest to the cruciform rod

2.   opposite the cruciform rod

3.   diagonal cells

Again, the results in each region can be divided into fuel, can and coolant. Clearly, the flux is lowest close to the cruciform rod, and a detailed absorption prediction by the CHARACTERISTICS method is compared with the best results in Table III.

Again, the results of the CHARACTERISTICS method lie between the "best" results obtained at AEEW and JAERI, even for the control rod absorption which is notoriously difficult to predict. Figure 13, reprinted from Halsall, shows the relationship between calculated eigenvalue and control rod absorptions for some of the submitted results.

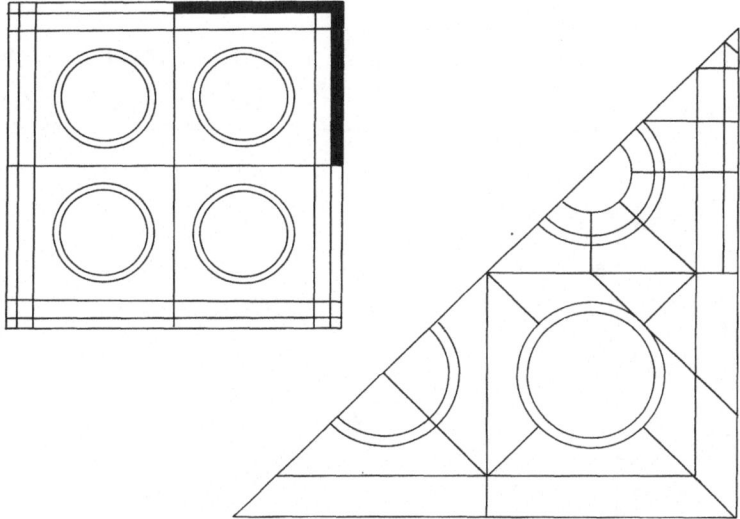

Figure 12. Geometry Used in Benchmark Problem 2

TABLE III

		WINF.	CHAR.	JAERI
OPPOSITE	FUEL	.1533	.1503	.1496
CRUCIFORM	CAN	.0020	.0020	.0019
	COOLANT	.0100	.0095	.0095
	FUEL	.3144	.3126	.3097
DIAGONAL	CAN	.0037	.0037	.0036
	COOLANT	.0170	.0169	.0167
CLOSEST	FUEL	.1305	.1320	.1321
TO	CAN	.0015	.0015	.0015
CRUCIFORM	COOLANT	.0055	.0055	.0056
CONTROL ROD		.3387	.3429	.3466
WATER GAP		.0237	.0232	.0231

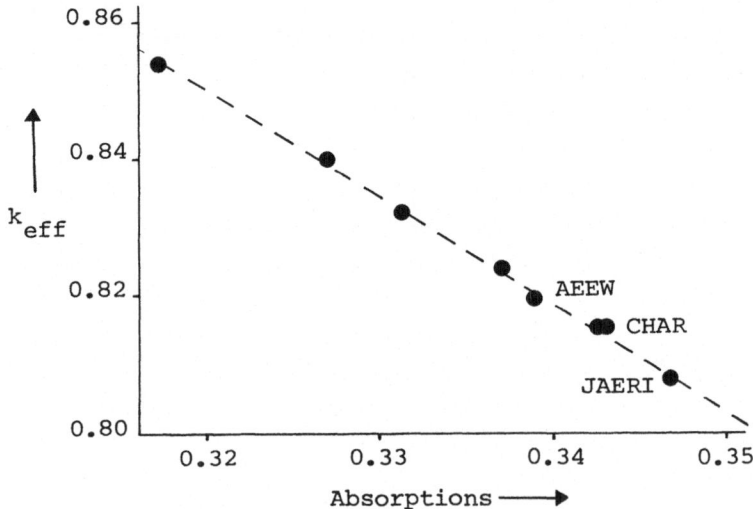

Figure 13. Effective Multiplication and Control-rod Absorption

V. FUTURE DEVELOPMENTS

With all methods of solving the transport equation, except in very simple problems, the use of computers is necessary. In recent years, the cost of programming and interpretation has become more important than that associated with hardware. Conceptually simple methods therefore will become "cheaper", when all things are taken into account, than more complex, possibly slightly quicker methods. The basis of this method is choosing some representative set of characteristics, together with boundary conditions and tracking along each to give mean angular fluxes, using Equation (2.8). Applying this idea to many different types of problems, we should bear in mind that a large amount of effort to make a computer code more "efficient" is not justified unless it makes the code very considerably quicker and does not obscure the underlying principle.

Wagner, Sargis and Cohen(14) have used a version of this method to solve for fluxes in a finite stack of cuboids. They used very special directions (perpendicular to faces and along diagonals), but tracking along more general directions should present little difficulty. If there are no incident neutrons on the stack, then we need only track once through the stack for each characteristic. Keeping a constant surface area density of characteristics and block size, the time taken for each inner iteration will be proportional to the stacks' volume and the number of angles used. This will be, for large stacks, very much quicker than any collision probability method.

Problems where the neutron flux differs greatly from isotropy may be dealt with by using a greater number of characteristics in directions of interest. Shielding or deep penetration calculation problems may be solved by this means; so also can problems arising from experiments where a parallel beam of incident neutrons from a reactor port is used.

At present, solutions to the multigroup form of the transport equation are found by an inner-outer iteration strategy. The inner iteration is used to solve for the angular and spatial distribution within one group, with an outer iteration for between-group transfers and fission vector calculation. It is possible to model phase space in a different way so we use a set of characteristics to solve for fluxes at

all angles and energies in parallel. The j label used in,
for example, Equation 2.8, would then specify a group and
an angle. For each region i, $\bar{Q}_{ij}$ would be calculated from
$\bar{\phi}_{ij}$ by means of an angle-energy scatter matrix and solution
of (2.8) then corresponds to both inner and outer iteration.

Given the data structures (arrays) defining (1), successive regions through which a lines passes; (2) lengths
of segments; (3) volumes; and (4) cross-sections, Equations
(2.9) and (2.8) require very little coding (ends of angles
being denoted by markers; e.g., negative region numbers).
The bulk of the work of finding the required solution could
be carried out, therefore, on a very simple processor that
had access to this tracked data. It might be that true parallel or array processing could be used for this.

ACKNOWLEDGMENTS

We should like to acknowledge an enormous debt to
Dr. John Askew at AEE Winfrith for suggesting this problem
and for his unfailing support and advice during its solution.
We are also grateful to M. J. Halsall for permission to use
diagrams and tables from a published report. One of us (M. B.)
is grateful to AEE Winfrith and Bristol Polytechnic for financial support as a Research Assistant during the period in
which this work was carried out.

REFERENCES

1.   Williams, M. M. R., "Slowing Down and Thermalization
      of Neutrons," North Holland.

2.   Weinberg, A. M. and Wigner, W. P., "Physical Theory of
      Neutron Chain Reactors," University of Chicago, 1958.

3.   Carlson, B. and Bell, V. J., "Solution of the Transport
      Equation by the S_N Method," Proceedings of the Second
      Geneva Conference, Page 2386, 1958.

4.   Green, C., "The Winfrith DSN Program, Mark 2." AEEW-
      R498, 1967.

5.   Askew, J. R. and Brissenden, R. J., "Some Improvements
     in the DSN Method of B. Carlson for Solving the Neutron
     Transport Equation," AEEW-R161, 1963.

6.   Askew, J. R., "Some Boundary Condition Problems Arising
     in the Application of Collision Probability Methods,"
     Proceedings of the Vienna Seminar, Page 343, 1972.

7.   Brough, M. D., "The CACTUS Code," (to be published).

8.   Askew, J. R., private communication.

9.   Meneghetti, D., "Discrete Ordinate Quadratures for Thin
     Slab Cells," Nuclear Science Eng., 14, Page 295, 1962.

10. Askew, J. R., "A Characteristics Formulation of the
     Neutron Transport Equation in Complicated Geometries,"
     Atomic Energy Establishment, Winfrith Report M1108,
     1972.

11. Brissendon, R. J., "A Formula for the Escape Probability
     of a Rod in a Uniform Lattice," Atomic Energy Estab-
     lishment, Winfrith Report R-282, 1963.

12. Mordant, M., "ZEPHYR, a Code for Solving Neutron Trans-
     port Problems on Irregular Meshes," Commissariat a
     l'Energie Atomique, B. P. 27, 1975.

13. Halsall, M. J., "Review of International Solutions to
     NEACRP Benchmark BWR Lattice Cell Problems," AEEW-
     R1052, 1977.

14. Wagner, M. R., Sargis, D. A., Cohen, S. C., "A Numer-
     ical Method for the Solution of Three-dimensional
     Neutron-transport Problems," Nuclear Science Eng., 41,
     1, pp 14-21, 1970.

HETEROGENEOUS CORE DESIGNS

FOR LIQUID METAL FAST BREEDER REACTORS*

P. W. Dickson and R. A. Doncals

Westinghouse Electric Corporation
Advanced Reactors Division
Madison, Pennsylvania, U. S. A.

I.   INTRODUCTION

Liquid Metal Fast Breeder Reactors (LMFBR's) are being developed in the United States, Russia, Great Britain, France, Germany and Japan. The use of LMFBR's extends the amount of energy that can be extracted from uranium by a factor of 60 to 70 as compared with light water reactors, thus converting a scarce energy source into a plentiful one; hence, the interest of the major nations of the world in LMFBR's.

The plutonium fuel in LMFBR's is normally in the form of mixed plutonium, uranium dioxides, and is contained in pellets within fuel rods separated by some mechanism to allow coolant flow. The rods are held in place within a hexagonal duct called a fuel assembly. Above and below the fuel pellets within the fuel rods are pellets of uranium dioxide. This uranium dioxide is called the axial blanket. Figure 1 shows the fuel assembly configuration. In most LMFBR's, the fuel assemblies are surrounded by similar assemblies containing rods filled with uranium dioxide as shown in Figure 2. These assemblies of uranium dioxide are referred to as the radial blankets. The purpose of the radial and axial blankets is to capture, in uranium, the neutrons escaping the fueled region. This uranium capture generates plutonium and enhances the breeding that occurs in the fueled

* This work was supported by DOE under Contracts EY-76-C-15-2395 and EY-76-C-02-3045M.

33

region.   The power in the blankets is lower than in the
fuel.   Therefore, the blanket rods can be and are, much
larger in diameter than the fuel rods.

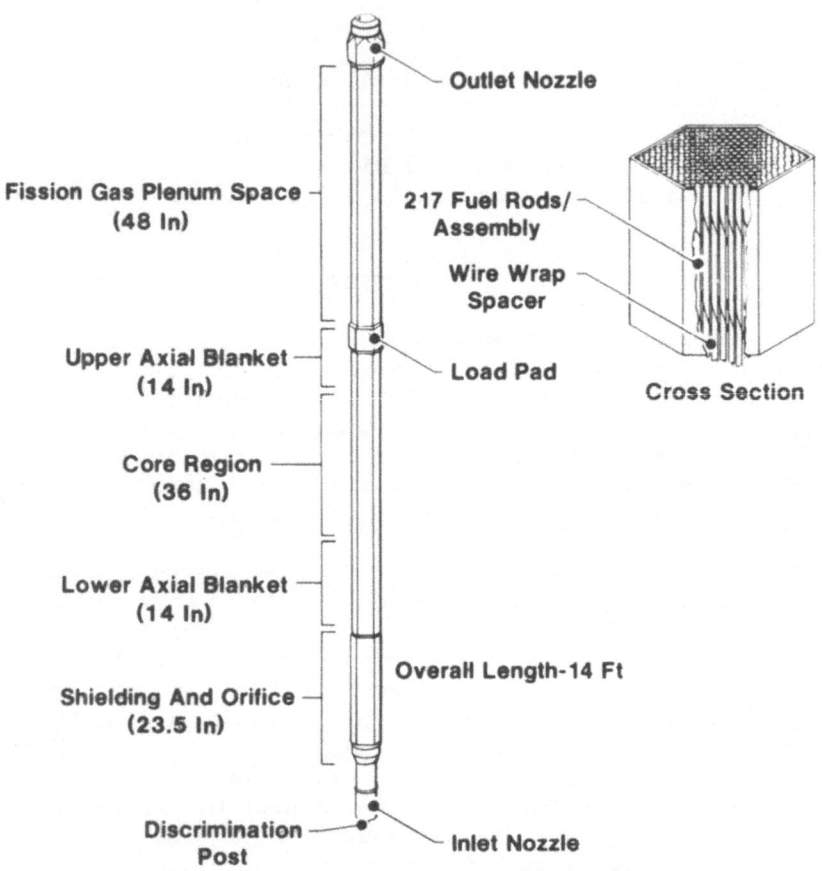

Figure 1.   Typical LMFBR Fuel Assembly

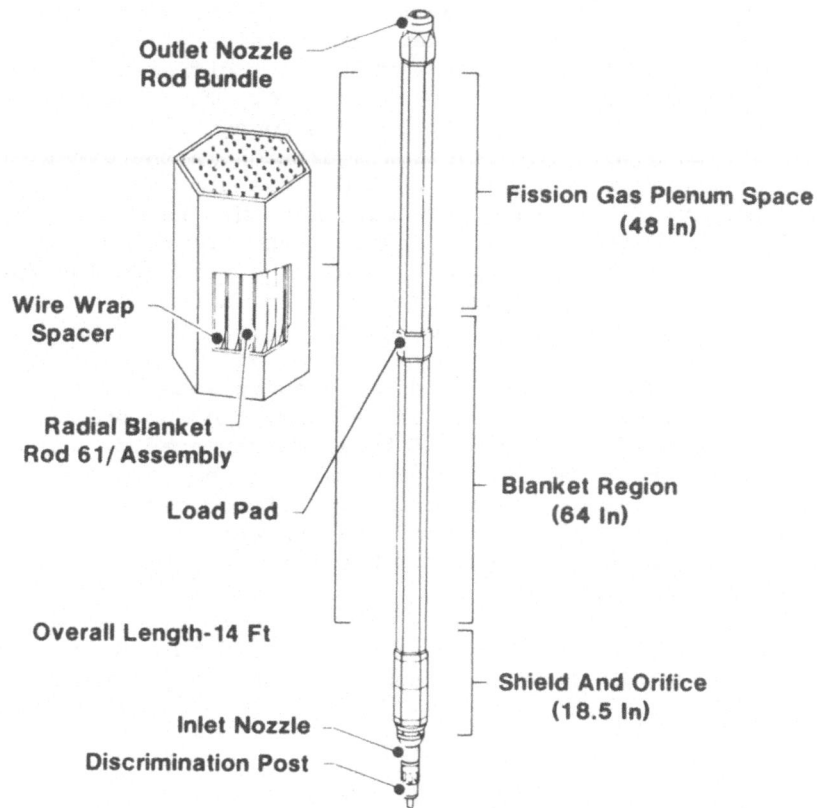

Figure 2. Typical LMFBR Radial Blanket Assembly

     To increase the uranium-238 fraction in the fueled
region and thereby increase breeding, the diameter of the
fuel rods can be increased at the sacrifice of coolant or
structural volume fraction. Alternately, blanket assemblies, with their large-diameter rods of uranium-238, can
be dispersed among the fuel assemblies. The result is called
a heterogeneous core, or more specifically, a radial heterogeneous core. The resulting characteristics are such that
the use of heterogeneous cores is being considered seriously
by designers in the United States, France and Germany, while
design studies have progressed in Great Britain.

In addition to the radial heterogeneous core, the concept of separating the fuel axially by insertion of blanket
pellets within the fuel rods has been considered(1,2,3).
This technique has advantages similar, but different in magnitude from the radial heterogeneous core. The disadvantages are a longer fuel assembly and a concern with fuel
slumping into a more reactive configuration in the event of
a melting accident. The merits of designs utilizing both
radial and axial heterogeneous zones have not been fully ex-
plored in the United States. However, the French have pursued studies of both radial and axial heterogeneous (parfait)
cores, and seem to feel that the axial parfait is a worth-
while addition to a reactor design(4).

This article discusses radial heterogeneous configurations, the reasons for considering them, the evolution of
the design concept, and some of their differences from homogeneous reactors. The latter portion describes two specific
examples, the flexibility afforded by heterogeneous designs,
the experimental verification of the concept, and its application in the event that fuels other than plutonium are to
be used in breeders.

Finally, the authors' ideas as to the future of heterogeneous designs in commercial applications are presented.

II. HISTORY OF LMFBR

The principal advantage of a liquid metal fast breeder
reactor is that it can be designed to breed more fissionable
material than it consumes. The boiling point of liquid metals also permits operation at high temperatures without
pressurization. In addition, the excellent heat transfer
properties of liquid metals permit the design of a smaller
core than would be possible with water, gas or organic coolants. These advantages motivated many of the leaders in
nuclear technology to initiate the development of this concept in the 1940's.

In spite of this early emphasis on sodium-cooled fast
reactors, their development and use as electrical power-
generating reactors has fallen far behind the development
of the slightly enriched water-moderated reactors. Instead,
fast breeder development has concentrated on developing the

technology, with only a few test or experimental reactors
being built. A synopsis and progression of the major liquid
metal fast breeder reactors that either have been built or
are under development in the United States are presented
here.

The Experimental Breeder Reactor No. I (EBR-I) was the
first reactor built in the United States to examine the
breeding and safety performance of a fast reactor using
sodium-potassium liquid metal as the coolant. The Argonne
National Laboratory designed and operated(5) the reactor
for approximately 13 years (from 1951 to 1963). Four cores
(Mark I, II, III and IV) were employed in the reactor
throughout its operating lifetime. Uranium-235 was utilized
as the fissile material in the Mark I through III cores, and
plutonium in Mark IV. The reactor had a thermal power rating of approximately one megawatt. The knowledge gained
from the tests performed gave considerable insight into the
proper design of fuel assemblies and core structures in a
fast reactor to minimize or eliminate their adverse effect
on the power coefficient. The breeding measurements provided the first real confirmation of the breeding performance in a liquid metal fast breeder reactor.

The Experimental Breeder Reactor No. II (EBR-II)(6)
located at the National Reactor Testing Station in Idaho
went critical in 1963 and is still currently in operation.
It was originally designed to be the first fast reactor to
operate with a closed fuel cycle. After successful demonstration of its original objective, it subsequently was
modified to become an irradiation test facility. Its thermal power rating has been increased to approximately 62 MW.
Irradiation data provided from this facility have been used
in the design of the Fast Flux Test Facility and currently
are being employed in the design of Clinch River Breeder
Reactor Plant.

The Enrico Fermi reactor was designed and built by the
Power Reactor Development Company to supply electricity to
the Detroit Edison System. It was located approximately 30
miles from Detroit, Michigan on the shore of Lake Erie. The
reactor plant was designed to use metal fuel and was rated
at 430 MWt; however, the first core was designed to produce
only 200 MWt. Initial criticality was achieved in 1963, and
all subsequent operation(7,8) was at less than 1 MWt until

the early part of 1966. At that time, a test program was
initiated for high power operation up to 100 MWt. A 60-
hour sustained operation at 100 MWt was achieved during
this period. However, on October 5, 1966, there was a partial meltdown of two fuel subassemblies during an ascent to
high power. Subsequent investigations showed that a segment
of the conical flow guide had become detached and blocked
coolant flow to four fuel subassemblies, two of which partially melted. Remote operating tools and instrumentation
were subsequently developed and used to remove the detached
segment and the other remaining segments. During this shutdown period, the tube-to-tube sheet joints of the steam generators were also modified by the use of an integral bore
weld to minimize leakage. In July, 1970, after completion
of all repairs and a complete assessment of the problem
leading to the test failure, the reactor again achieved
criticality and operated successfully until 1972. In November, 1972, the Power Reactor Development Company decided to
decommission the reactor because it could not fund a proposed
follow-on program for operation with an oxide core.

The Southwest Experimental Fast Oxide Reactor(9,10)
(SEFOR) was designed to demonstrate the inherent safety
characteristics of a fast reactor using oxide fuel and sodium coolant. SEFOR was funded jointly by the Karlsruhe
Laboratory of the Federal Republic of Germany, Euratom,
SAEA (Southwest Atomic Energy Associates, a group of utilities) and the General Electric Company. Special emphasis
was placed on obtaining a good understanding of the Doppler
coefficient. The initial core loading was completed in
April, 1969. The Doppler tests performed were in good agreement with analytical prediction, thus providing considerable
confidence to corresponding power reactor analysis. After
a successful test series, the reactor was shut down in January, 1972, and subsequently decommissioned.

The Fast Test Reactor(11) is a sodium-cooled reactor
presently being built on the Hanford Reservation in Washington. Westinghouse Hanford Corporation is the prime contrac-
tor to the Department of Energy for management and operation
of the facility. The Westinghouse Advanced Reactors Division was the lead reactor designer. The reactor provides
a well-instrumented environment for testing material specimens and assemblies of fuel, blanket and control rods. Test
assemblies can be positioned within the core to be cooled by
the reactor primary coolant. Alternately, closed loops are

provided with separate coolant flow for specimen irradiation under prescribed conditions of temperature, pressure and flow. Initial criticality is expected in late 1979. The reactor is designed to produce 400 MWt of power that will be discharged to the atmosphere.

The Clinch River Breeder Reactor Plant(12) (CRBRP) is a combined government and industry effort to build the first fast breeder reactor demonstration plant in the United States. The plant will be constructed in the Roane County portion of Oak Ridge, Tennessee. It has an electrical power rating of 350 to 400 MWe that will be utilized in the TVA power grid. Overall management is achieved through an integrated organization of the Department of Energy, industry personnel and the major project partners (Tennessee Valley Authority and the Commonwealth Edison Corporation of Chicago). The major industry participants are the Westinghouse Electric Corporation, Burns and Roe, Inc., Stone and Webster Engineering Corporation, General Electric Corporation, Atomics International and various component manufacturers. Final design is complete or approaching completion for all major reactor components. Many components presently are being manufactured. The present target date for initial criticality is in the mid-1980's.

Liquid metal fast breeder reactors also are being developed in France, Germany, Japan, the United Kingdom and the Union of Soviet Socialist Republics. Table I shows the reactors presently being operated by their respective countries and those planned for commercial operation in the near future. Six reactors presently are in operation, providing their countries with data for extrapolation to larger sizes. These initial reactors have power ratings up to 350 MWe. As indicated in Table I, the second generation of reactors have power ratings of 600 to 1200 MWe and are quite suitable for deployment as commercial reactors. Although the French have performed considerable work with the heterogeneous concept, they have not made definite plans to date for incorporation of the concept into their operating or planned reactors. Similarly, reactors developed or under development in the United States all utilize the homogeneous concept. These reactors consist of fuel assemblies surrounded by either reflectors and/or blanket assemblies. It was not until recently that any special design studies had been performed to evaluate the heterogeneous concept that employs fertile assemblies interspersed among the fuel assemblies.

P. W. DICKSON AND R. A. DONCALS

TABLE I

FOREIGN LMFBR REACTORS [1]

Country	Reactor	Power		Actual/Scheduled	
		MWt	MWe	Init. Critical	Full Power
France	Rapsodie	40	--	1966	1967
	Marcoule Phenix	563	250	1973	1974
	Super Phenix	3000	1200	1979	1980
Germany	Karlsruhe KNK	58	21	1971	1972
	Kalkar SNR	762	327	1981	1982
Italy	Brasimone PEC	130	--	1978	1978
Japan	Joyo	50	--	1977	1978
	Monju	714	300	1984	1984
United Kingdom	Dounreay DFR	60	15	1959	1963
	PFR Dounreay	600	270	1974	1976
Union of Soviet Socialist Republics	Obninsk BR5	--	5	*	*
	Melekess BOR-60	60	12	1969	1971**
	Shevchenko BN350	1000	350	1972	1973
	BN 600	1500	600	1978	1978

* Prior to 1959.

** Scheduled, but Actual Dates Not Available.

[1]From "Technical Data" and "Progress Schedules", "Power Reactors 1977",

Nuclear Engineering International, April 1977 Supplement.

III.   EVOLUTION OF THE HETEROGENEOUS DESIGN CONCEPT

The evolution of the heterogeneous reactor design concept in the United States actually started with the design and operating experience gained from the Shippingport reactor. This system is a pressurized water reactor (PWR), cooled and moderated by light water. The initial core used the "Seed and Blanket Concept"(13,14,15) which was developed at the Bettis Atomic Power Laboratory and Knolls Atomics Power Laboratory for Naval Reactors. In this concept, an annular seed region of highly enriched U-235 is surrounded by natural uranium blankets. Control is achieved by the use of movable control rods.

The reported advantages of this concept are:

1.   reduction in investment of enriched
     fissionable material

2.   generation of a substantial portion
     of the core power from natural uranium

3.   satisfactory negative temperature
     coefficients

4.   reduced mechanical control requirements

The above advantages are enhanced for larger power systems. In addition, the development of a variable seed geometry(16) provides a unique method of control that avoids parasitic capture of neutrons. In this scheme, the seed is made in two concentric cylinders. Movement of one of the concentric cylinders makes the effective seed region thinner and longer, and increases the neutron leakage to the blankets to reduce core reactivity. This method of movable seed and blanket control is used in the Light Water Breeder Reactor.

In the fast reactor field, heterogeneous concepts were first introduced by two of the four participants in a study (17,18,19,20) initiated by the U. S. Atomic Energy Commission* for input to their long-range planning program. The Westinghouse design(17) consisted of seven identical modules. Each module consisted of a central fuel region within a radial blanket, surrounded by a mechanical and thermal barrier of graphite. This modular arrangement enhanced the breeding by exposing most of the radial blankets

---

* See footnotes, p.84

to the neutron leakage from the fuel assemblies. The Allis
Chalmers heterogeneous design(18) was an annular core
approximately four feet high and two feet thick. Outside
the inner and outer diameters of the core were 18-inch thick
radial blankets. Axial blankets also were utilized. The
primary incentive for these heterogeneous concepts was to
reduce sodium void worth, while still maintaining high
breeding and performance capability.

The complexities and overall size of these early designs
tended to be detrimental to the heterogeneous concept. Consequently, follow-up studies, such as those performed by
the General Electric Corporation(21) and the Westinghouse
Electric Corporation(22), concentrated on the design of the
homogeneous core configuration. The design emphasis shifted
away from the heterogeneous concept and the potential benefit of its reduced sodium void worth. The proposals submitted to the Project Management Corporation and the Atomic
Energy Commission for the first LMFBR Demonstration Plant
(Clinch River Breeder Reactor Plant) utilized the homogeneous core configuration as have all fast breeder reactors
built in other nations.

Studies continued to be performed at the Massachusetts
Institute of Technology and subsequent follow-on work by
the Electric Power Research Institute(3) to evaluate the
axial heterogeneous concept. In addition, a seed-blanket
modular concept also was evaluated(23). However, not until
the publication(24) of analysis by the French in 1975 was
there renewed interest in the radial heterogeneous concept
in the United States by several reactor manufacturers and
the National Laboratories. The stated merits of the concept were:

1. reduced doubling time

2. decreased sodium void coefficient

3. reduced core fluences

4. flatter power distribution (which
    permits the use of a single fuel
    enrichment without excessive radial
    power peaking)

Subsequently, a design study performed by the Westinghouse Electric Corporation(25) showed the feasibility and
applicability of this concept to the CRBRP, not only neutronically, but also regarding all other engineering as-
pects(26). These studies also indicated:

1. a significant increase in breeding gain
    (0.11) and a large reduction in doubling
    time

2. reduced ($\sim$25%) neutron fluences

3. reduced ($\sim$50%) sodium voiding worth
    in the fuel assemblies (usually
    referred to as fuel Na void coefficient)

4. the ability to design with a single
    fuel enrichment

The single most detrimental characteristic was that the
fuel inventory was increased by approximately 25%. However,
the resulting enrichment was approximately the same as utilized in the outer core zone of the homogeneous concept during the equilibrium cycle. In addition, subsequent anal-
ysis(27) performed at the Argonne National Laboratory showed
that this heterogeneous CRBRP had an improved response to
hypothetical accident transients not terminated by the plant
protection system. This benefit was derived from the temporal incoherence between boiling in the fuel and inner
blanket assemblies, as well as the lower sodium void coefficient.

Based on these favorable results, the applicability of
the heterogeneous concept for large commercial reactors
has been evaluated by the Argonne National Laboratory(28,
29,30) and the United States reactor manufacturers(31,32,33).
In addition, considerable work(34,35,36,37) has and is
presently being performed by European nations. These domestic and European studies show general agreement relative
to the merits and disadvantages of the concept, with some
exceptions noted later. Considerable effort is still under-
way.

One reason the French favor the heterogeneous concept
is that it provides a means of using Phenix-tested fuel pins

in their larger commercial designs. A heterogeneous design
provides many of the same effects as would the use of a larger fuel pin diameter without the development costs associated with developing fuel with a larger pin diameter. In
the United States, testing of larger pin diameters already
has been initiated for commercial reactors. Consequently,
the above positive merit will not be as important in the
final selection of reactor concept in this country. The
domestic discussions or debates are related more to the
necessity of reducing the effective sodium void by the use
of the heterogeneous concept to ensure a benign hypothetical
core disruptive accident versus the penalty associated with
higher fissile loadings.

        IV. BREEDING RATIO, DOUBLING TIME AND
    SODIUM VOID CHARACTERISTICS OF HETEROGENEOUS CORES

     The major incentives for considering heterogeneous cores
derive from their increased breeding ratio, reduced sodium
void reactivity worth and reduced hypothetical disruptive
accident energetics. Additional benefits are the reduced
sensitivity of doubling time to fuel rod diameter as compared
with a homogeneous core (this is particularly significant
because retaining a proven fuel design enhances ability to
license), a reduction in the number of required fuel enrichment zones, a higher power/fluence ratio, a greater percentage of power produced in blanket assemblies, and generally,
an improved peak/average power ratio. These latter advantages are discussed further in Section VI, C.

     For a given fuel assembly design, a heterogeneous con-
figuration typically will increase the breeding ratio by
approximately 0.1(25) relative to a homogeneous design of
the same power level. This gain in breeding is primarily the
result of the introduction of the greater fertile mass of the
blanket assemblies and its effect on the neutron importance
in the reactor. The neutron spectrum in a heterogeneous design has a higher average energy that tends to improve breeding.

     Figure 3 shows the start of equilibrium cycle fissile
mass, breeding ratio and doubling time as a function of the
fuel volume fraction for a fixed 1,000 MWe design. This

particular heterogeneous design is identified as "Design A"
in Section V, where more detailed technical data are presented. It is discussed here to indicate general trends. An
increase in fuel volume fraction results in an increase in
both the fissile mass requirement and the breeding ratio.
The difference in the signs of the slopes of these two curves
results in a minimum in the doubling time curve. The heterogeneous design has a broad minimum ranging between fuel volume fractions of 40 and 50%. The homogeneous doubling time
curve was calculated(38) for a larger-size reactor and ex-
trapolated to the smaller plant. Of particular interest is
that the homogeneous design has a much sharper doubling time
minimum, and as a result, is more sensitive to the fuel volume fraction. The second point of interest is that both designs have minimums for the same fuel volume fractions of
0.425 to 0.45. The third point is that the heterogeneous
design has approximately a two-year doubling time advantage
despite its higher initial fissile investment.

   The fuel sodium void reactivity is smaller in a heterogeneous core than in a homogeneous core by roughly a factor
of two, even when the heterogeneous core has been designed
for good performance, independent of sodium void reactivity
considerations. The fuel sodium void worth is reduced in any
heterogeneous design by the reduced number of fuel assemblies
in the high central positive worth region, the higher fissile
loading that results in a reduced positive component of sodium void reactivity and the increased neutron leakage into
the blanket assemblies. The sodium void reactivity can be
reduced further by thick rings of blanket assemblies separating rings of fuel(39). The thicker the rings of blanket
assemblies, the lower is the sodium void reactivity. These
thick blanket regions tend to neutronically decouple the fuel
regions, and the thicker the rings, the greater is the degree
of decoupling. Decoupling means that each ring of fuel tends
to behave more as an independent reactor, requiring separate
control, independent of control rods in other rings.

   Greater decoupling reduces core performance. Consequently, the design of decoupled cores is a trade-off of allowable sodium void reactivity and core performance. Virtually
all designs today are of tightly-coupled cores, with sodium
void reactivities on the order of $2 to $3, but with good
performance.

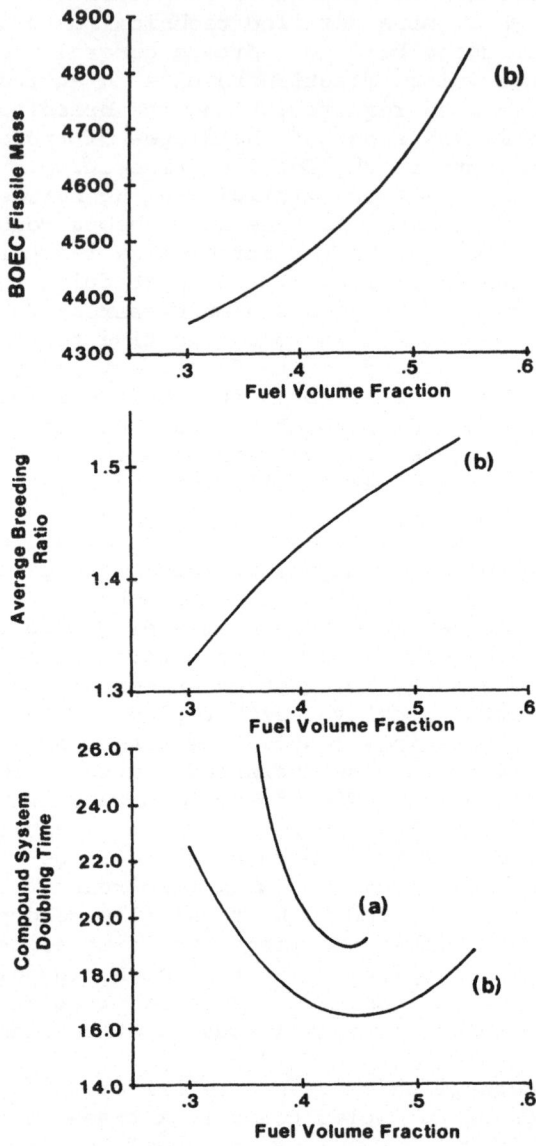

Figure 3.   Dependence of Fissele Mass, Breeding Ratio and
            Doubling Time on Fuel Volume Fraction: (a),
            1,000 MWe Homogeneous Design; (b), 1,000 MWe
            Heterogeneous Design

Another important merit of the heterogeneous design is that it gives the designer another degree of freedom for flattening the radial power distribution. In a large homogeneous core, two to three fuel enrichments are employed to flatten the radial power distribution. In principal, homogeneous cores could be designed with different fuel enrichments azimuthally, such as higher enrichments near control rods. However, the extra complexity of design has precluded azimuthal zoning in homogeneous designs. In the heterogeneous concept, the designer has the ability to arrange the fuel and inner blankets to minimize the power peaking, both azimuthally and radially, while employing only one fuel enrichment. This extra degree of freedom complicates core optimization because the fuel/blanket pattern also must be compatible with control, breeding, fissile inventory and possibly sodium void requirements. However, it is possible to satisfy all these requirements with a single fuel enrichment in a heterogeneous design.

V. DESIGN OF HETEROGENEOUS REACTORS

Numerous heterogeneous reactor designs have been generated by many organizations. The design that has been studied in the most depth in the United States, and quite possibly the world, is the design of a heterogeneous configuration for the Clinch River Breeder Reactor utilizing the reference oxide fuel design; i.e., the fuel design for the initial core of the homogeneous version of the CRBRP. A similar preliminary evaluation also has been performed for a prototype large breeder reactor.

A. Clinch River Breeder Reactor Plant

The CRBRP study (25) was performed to determine the feasibility of replacing the homogeneous core arrangement with a heterogeneous core without violating the physical interface of core barrel size and control rod locations, or the system constraints of fluence limits, power, refueling interval or safety characteristics. The resulting heterogeneous core arrangement is compared in Figure 4 with the reference homogeneous configuration. This heterogeneous design was the result of considerable studies performed to achieve, with a single fuel enrichment, the required reactivity worth in both the primary and secondary shutdown sys-

tems without exceeding peak linear power limitations in the
fuel and blanket assemblies, while achieving a specified
breeding ratio.

The homogeneous core has 198 fuel assemblies that are
subdivided into two enrichment zones to minimize the radial
power-peaking factor. The fuel zone is surrounded by approximately 2 1/2 rows of radial blankets and four rows of re-
movable shield assemblies employing Inconel rods in a stainless steel duct.

The fuel assembly structural member is a hexagonal
stainless steel duct containing 217 fuel rods in a triangular
array. Each rod is 0.23 inch in diameter, and has a cladding
thickness of 15 mils. The blanket assemblies are of the same
hexagonal cross-section, except that they contain only 61
rods, with a 0.506-inch outer diameter. The height of the
active fuel region is 36 inches, bounded by 14-inch axial
blankets containing depleted uranium. The equilibrium fuel
management scheme in the homogeneous design employs scatter
refueling of the fuel and shuffling of radial blankets.

The fuel, blanket, control and shield assemblies used
in the homogeneous core are mechanically identical to those
employed in the homogeneous concept except for minor changes
in the inlet end hardware. In addition, the inner and radial blankets are identical in design except for inlet nozzles and discriminators. The heterogeneous configuration
consists of 156 fuel assemblies with annular rings of inner
blankets (broken by fuel "islands" surrounding the primary
outer-ring control rods), surrounded by two rows of radial
blankets. The design contains a total of 15 control locations, 9 primary control assemblies (six row-7-corner locations and three at alternating row-4-corner locations) and
6 secondary control assemblies (in the six row-7-flat locations). The result is a 15-control-rod heterogeneous design
as opposed to the 19 control assemblies used in the homogeneous concept.

In the heterogeneous design, complete refueling of fuel
and inner blanket assemblies is performed every two years,
at the end of even cycles. A partial refueling is required
at the end of each odd cycle. This partial refueling consists of replacement of six inner blanket assemblies in row-
5-corner positions with fuel assemblies for reactivity en-

hancement. This unusual refueling sequence is necessary in
CRBRP because of the constraints on control rod locations,
refueling interval and fuel design, but it would not be preferred for a new reactor design.

The peak calculated linear power ratings for the homogeneous and heterogeneous designs are 14.4 and 15.9 kW/ft.
The homogeneous design had a limit of 16 kW/ft, but did not
approach the limit.

Annual refueling is planned for all operating cycles.
All fuel and inner blanket assemblies are replaced at the
beginning of odd-numbered cycles and the six blanket assemblies in row-5-corner positions are replaced by fuel assemblies at the start of even-numbered cycles in order to make
up the reactivity excess for the forthcoming burnup cycle.
Beginning-of-cycle 2 is an exception in that only three
alternating row-5-corner positions are refueled. The core
fuel assemblies contain a single enrichment of low Pu-240
plutonium that varies from 33.4 to 34.0% in the first five
years of operation.

The initial core, and all subsequent odd-numbered cycles,
will contain 156 fuel assemblies, 82 inner blanket assemblies
and 132 radial blanket assemblies. The first operating cycle
is planned for 128 full power days (fpd) over a one-year interval, allowing for reduced operation during initial plant
testing. Three inner blanket assemblies, at alternating row-
5-corner positions, 60 degrees removed from the row-4-control
channels, will then be replaced by three fuel assemblies for
the second cycle of 200 fpd over a one-year interval. At the
end of this second cycle, all fuel and inner blanket assemblies are discharged, while all radial blanket assemblies
remain in place. After the third cycle of 275 fpd (one year),
the six inner blankets at the row-5-corner positions are replaced by six fresh fuel assemblies for the fourth cycle of
275 fpd. At the end of cycle 4, all fuel and inner blanket
assemblies are again discharged. The fuel and inner blanket
management schemes for cycles 3-4 are repeated for every subsequent two cycles. At the end of cycle 4 and every four
cycles thereafter, the entire first row of radial blanket
assemblies is replaced. At the end of cycle 5 and every subsequent five cycles, the outer row of radial blanket assemblies is replaced. The radial blanket assemblies are not
shuffled as they are in the fuel management scheme of the
homogeneous design.

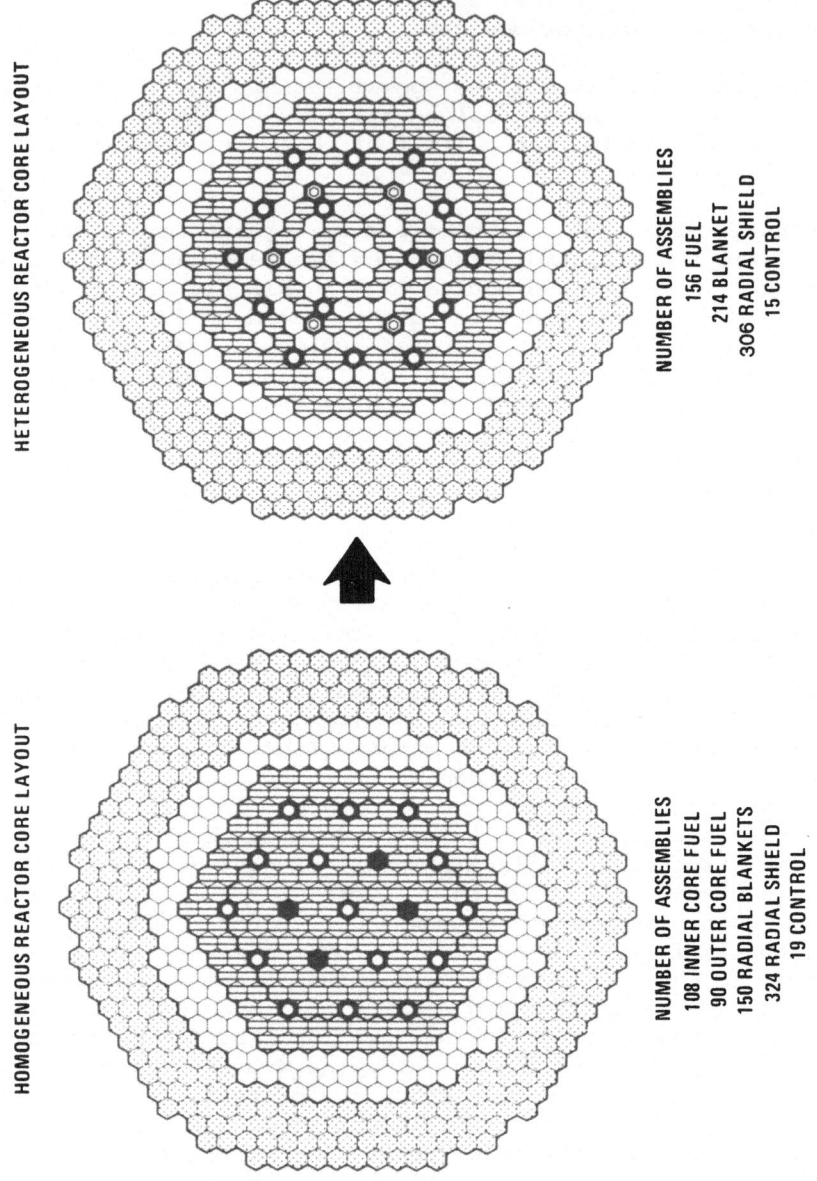

HETEROGENEOUS REACTOR CORE LAYOUT

NUMBER OF ASSEMBLIES
156 FUEL
214 BLANKET
306 RADIAL SHIELD
15 CONTROL

HOMOGENEOUS REACTOR CORE LAYOUT

NUMBER OF ASSEMBLIES
108 INNER CORE FUEL
90 OUTER CORE FUEL
150 RADIAL BLANKETS
324 RADIAL SHIELD
19 CONTROL

Figure 4.   Comparison of 1,000 MWt Homogeneous and Heterogeneous Reactor Core Layouts

The core fuel assemblies, which are identical to those used in the homogeneous design, employ a single fuel enrichment and mixed uranium-plutonium oxide fuel pellets. The feed enrichment at the beginning-of-cycle one is set according to the reactivity necessary to guarantee hot-full power criticality at the end of cycles one and two. The reactivity swing due to fuel depletion during the cycle is the main contributor to the enrichment requirement. Other reactivity effects that contribute to the required excess margin are uncertainties in power defect, fissile content, criticality, and burnup reactivity swing. The resulting plutonium enrichment required is 33.4%. The fissile loadings for start-of-cycles 3 and 5 are 1525 kg and 1552 kg, respectively.

Figure 5 illustrates the breeding ratio during the first six cycles for both the homogeneous and heterogeneous designs using plutonium with an isotopic composition representative of that discharged from light water reactors. The difference in the equilibrium values (cycles 4 to 6) indicates that the heterogeneous core concept increases the breeding ratio by 0.11 over that of the homogeneous design.

Figure 6 shows the relative merits of this increased breeding for the heterogeneous design in the form of a comparison of net fissile plutonium required versus the year of operation of the plant. For the homogeneous core, the initial plutonium investment is approximately 1250 kg, increases to a peak at 1400 kg in the fourth and fifth years and decreases to 875 kg at the end of 30 years. At that time there are 1550 kg of plutonium in the fuel and blankets. The net fissile gain for this core over the 30 years of operation is 675 kg (1550 minus 875) after the reactor is decommissioned and its plutonium is returned to the stockpile. For the heterogeneous design, the initial peak fissile investment is slightly higher, 1600 kg, but due to its better breeding performance, the investment at the end of 30 years is only 350 kg. This results in a net fissile gain of 1465 kg (1815 kg in the fuel and blankets at the end of 30 years minus the 350 kg still owed to the stockpile), over twice that of the homogeneous concept.

The compound system doubling time (CSDT) is a measure of the potential growth rate of an LMFBR-based electric economy. As such, it is an indicator of breeding ratio (total reactor fissile invested, etc.)

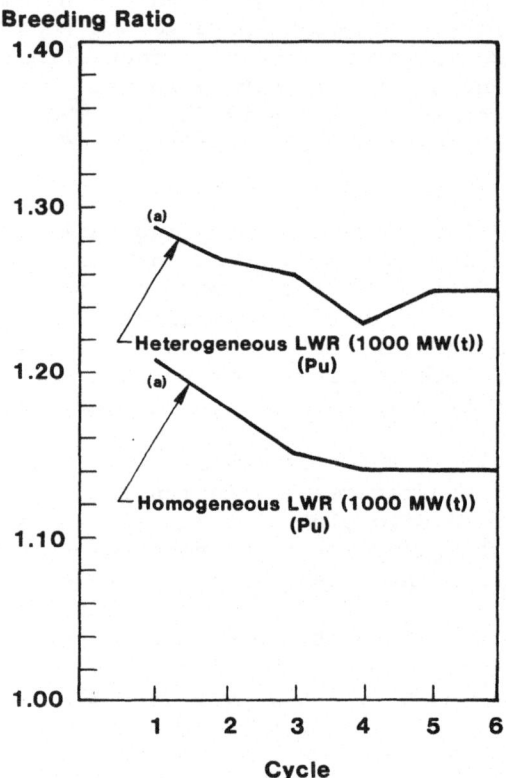

Figure 5.   Breeding Performance of Homogeneous and
Heterogeneous Cores

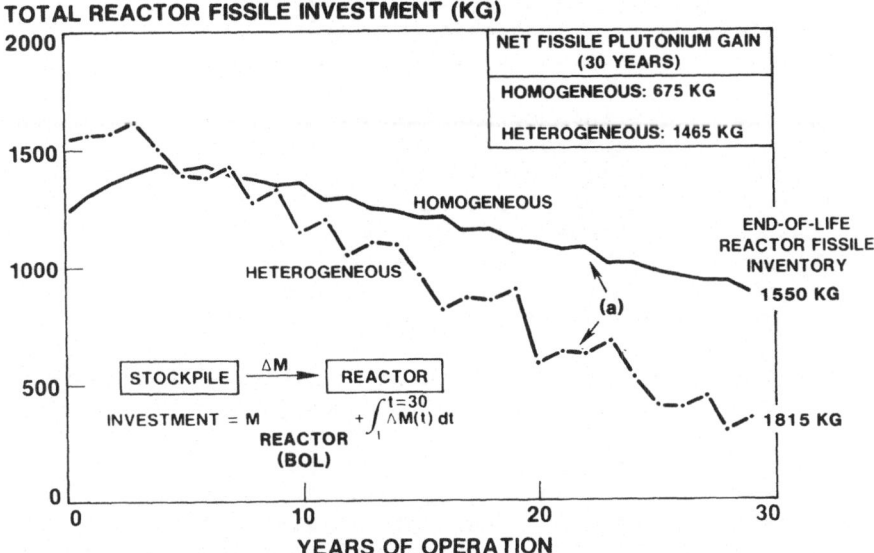

Figure 6. Net Fissile Plutonium Investment from Stockpile (Initial Fuel Assembly, Low Pu-240 Plutonium)

The compound system doubling time is defined according to

$$CSDT\ (yr) = \frac{\ln 2 * System\ Factor * Fissile\ Mass}{(Fissile\ Gain-Fissile\ Loss)*cycles/year}$$

where
System Factor corrects the reactor fissile inventory for required out-of-pile fissile material committed to the plant, generally assumed to be one reload core out-of-pile for one year.

Fissile Mass is the equilibrium beginning-of-cycle (BOC) fissile plutonium (Pu-239 + Pu-241) inventory in core, excluding that in the blankets.

Fissile Gain is the net plutonium fissile
mass (Pu-239 + Pu-241) gain per cycle,
including consideration of Pu-241 decay.

Fissile Losses include fabrication losses
(1/2% of one feed-reload batch), reprocessing losses (1/2% of one reload batch plus
1/2% of fissile gain) and Pu-241 decay
loss in the external cycle (4.6% decay of
the average Pu-241 inventory in the out-
of-pile batch per year plus 4.6% decay of
the average Pu-241 inventory in core).

Based on detailed mass inventories obtained from two-
dimensional burnup calculations, the homogeneous design
would have a compound system doubling time of 62 years with
light water reactor plutonium. The corresponding value for
the heterogeneous design is 33 years. This large difference
is due primarily to the larger net fissile gain per year for
the heterogeneous design which more than compensates for the
larger fissile mass.

In addition to its superior breeding performance, the
heterogeneous concept has a reduced fuel sodium void reactivity worth and the added feature of temporal incoherence
of voiding between the fuel and blanket assemblies (see
Table II). Values presented for fuel and inner blanket
voiding are for voiding of the 36-inch high section of the
assembly (excluding the 14-inch axial blankets). The total
reactor voiding also includes reactivity contributions from
the axial as well as the radial blankets. The change in
voiding worth with burnup reflects the flux redistribution
resulting from plutonium buildup in the blankets in the
inner regions of the core as well as the odd-year row-5 refueling. As indicated on the table, the maximum positive
fuel assembly voiding reactivity is only $2.26 in the heter-
ogeneous design compared with $4.0 for the homogeneous de-
sign. This results in a considerable reduction in the reactivity that can be postulated to be available during a hypothetical overpower transient with failure to scram. Because
the blanket assemblies are slightly overcooled and have a
longer time constant than the fuel, the hypothetical accident
would be terminated by fuel dispersion prior to significant

addition of sodium void reactivity from the blankets, while
the Doppler feedback available from the blankets tends to
reduce the peak reactivity insertion.

TABLE II

SODIUM VOID REACTIVITY ($)
1000 MW(t) Reactor Core Layout

	Heterogeneous		Homogeneous	
	BOC1	EOC4	BOEC	EOEC
o Fuel Assembly Voiding	-0.97	+1.50	+3.28	+3.34
o Inner Blanket Voiding	+0.88	+1.64	--	--
o Total Reactor Voiding	-1.56	+2.21	+2.45	+2.52
o Maximum Positive Voiding				
Fuel	+0.98	+2.26	+3.90	+4.00
Inner Blankets	+0.90	+1.64	--	--
Total	+1.88	+3.90	+3.90	+4.00

Table III compares the homogeneous and heterogeneous
Doppler constants for the beginning of life (BOC1) condition. The fast-acting fuel Doppler constant in the heterogeneous system essentially is half that of the homogeneous
concept. However, the contribution of the heterogeneous
inner blankets make the total Doppler constants for the two
systems essentially the same. In normal plant trips, the
difference in the fast-acting fuel Doppler constants has
only a very small effect on the maximum of the power rise.
During a hypothetical accident transient, the boiling sequence in the fuel and blankets is such as to reduce the already low probability of an energetic HCDA in both the
demonstration and commercial-size reactors.

TABLE III

$$1000 \text{ MW(t) DOPPLER CONSTANT, } -T\frac{dk}{dT} \times 10^4$$

Homogeneous		Heterogeneous	
	BOC1		BOC1
Inner Core	43.0	Core	23.2
Outer Core	12.9	Inner Blanket	32.1
Radial Blanket	7.0	Radial Blanket	14.5
Lower Axial Blanket	3.4	Lower Axial Blanket	1.8
Upper Axial Blanket	1.0	Upper Axial Blanket	0.8
Total	67.3	Total	72.4

B. Prototype Large Breeder Reactor (PLBR)

The applicability of the heterogeneous concept for large commercial reactors currently is being evaluated by the national laboratories and by various reactor manufacturers. This section summarizes a "Radial Parfait Core Design Study"(33) performed by the Westinghouse Electric Corporation for the United States Government and the Electric Power Research Institute. The primary objective of the study was to design a radial parfait core which, through its inherent characteristics, would give assurance that an energetic core disruptive event would not result from a transient overpower or undercooling accident without scram.

This comprehensive design study resulted in three promising PLBR candidates (see Figures 7, 8 and 9) that were identified as Designs A, B, and C. All of these designs used 252 fuel assemblies, 241 blanket assemblies and 18 control assemblies. Each design utilized an identical fuel assembly design with a four-foot core height and 271 rods (0.31-inch O. D.) per assembly. Identical 127-rod blanket assemblies and 37-rod control assemblies also were used. The upper and lower axial blanket was 14 inches thick.

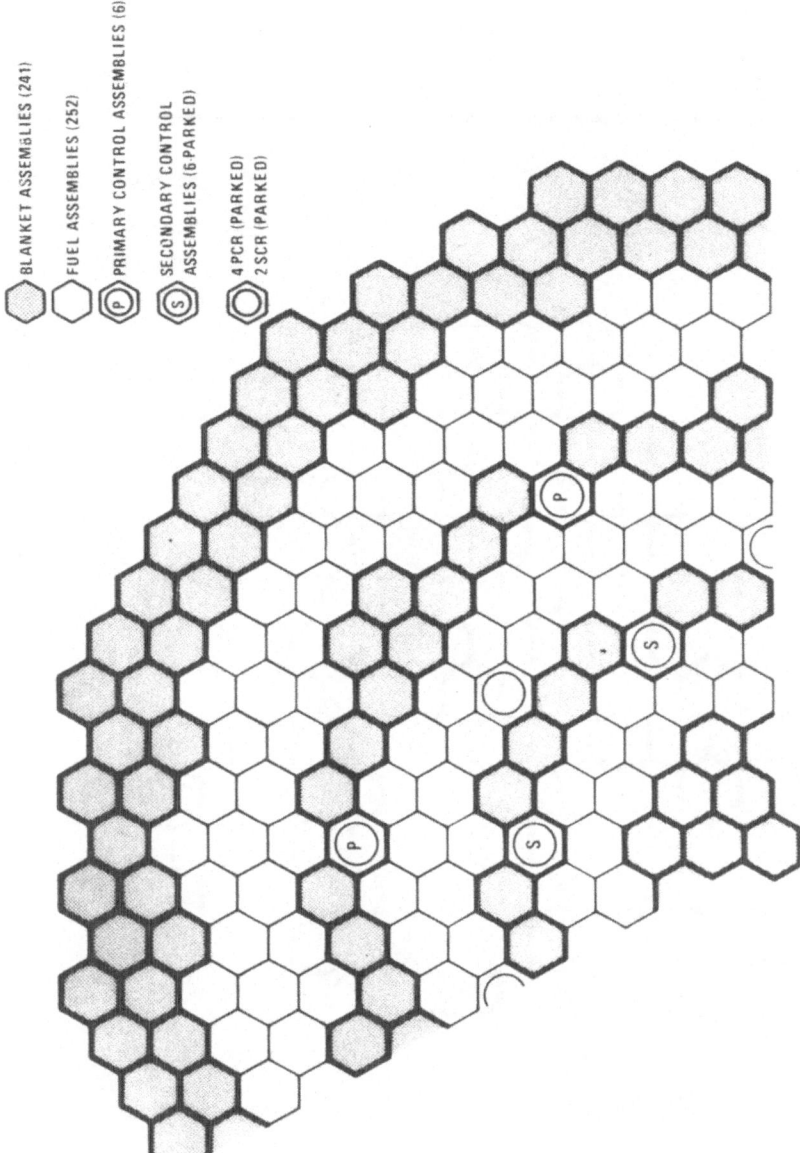

BLANKET ASSEMBLIES (241)

FUEL ASSEMBLIES (252)

(P) PRIMARY CONTROL ASSEMBLIES (6)

(S) SECONDARY CONTROL
ASSEMBLIES (6 PARKED)

4 PCR (PARKED)
2 SCR (PARKED)

Figure 7. Design A, Heterogeneous Core Layout, 2550 MWt

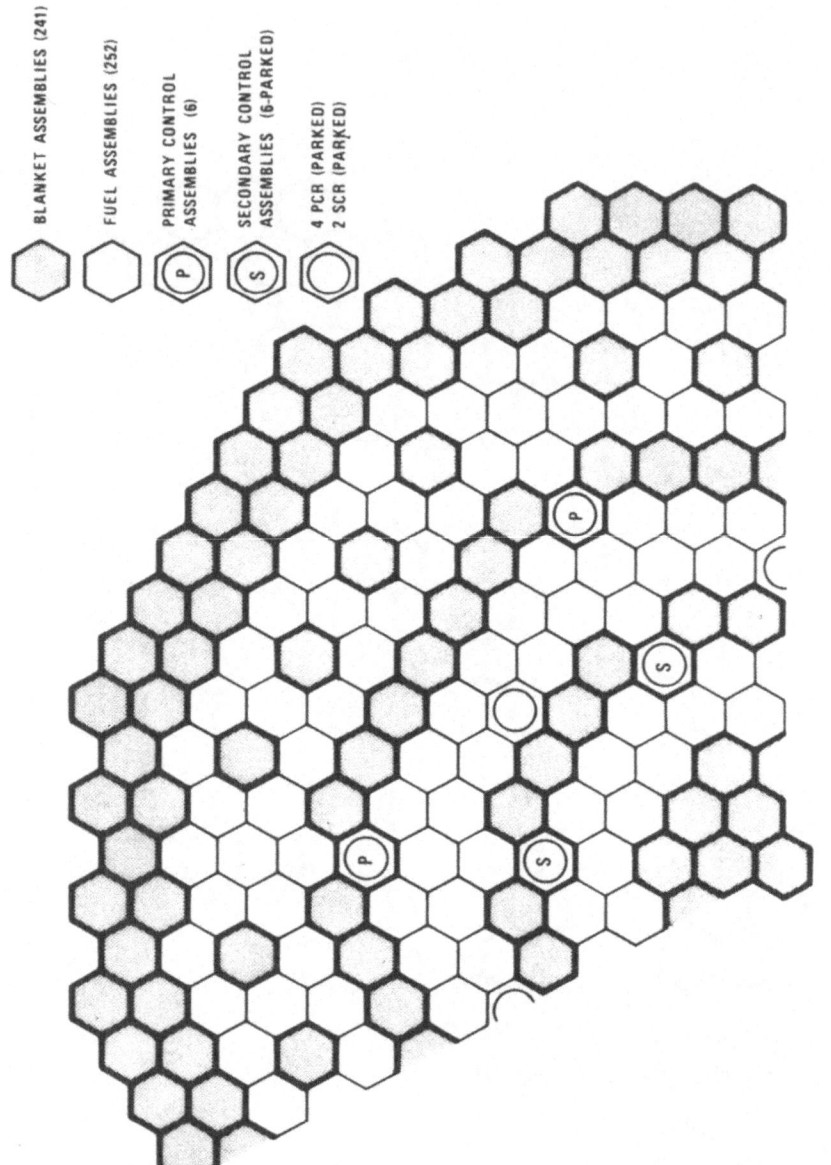

Figure 8. Design B, Heterogeneous Core Layout, 2550 MWt

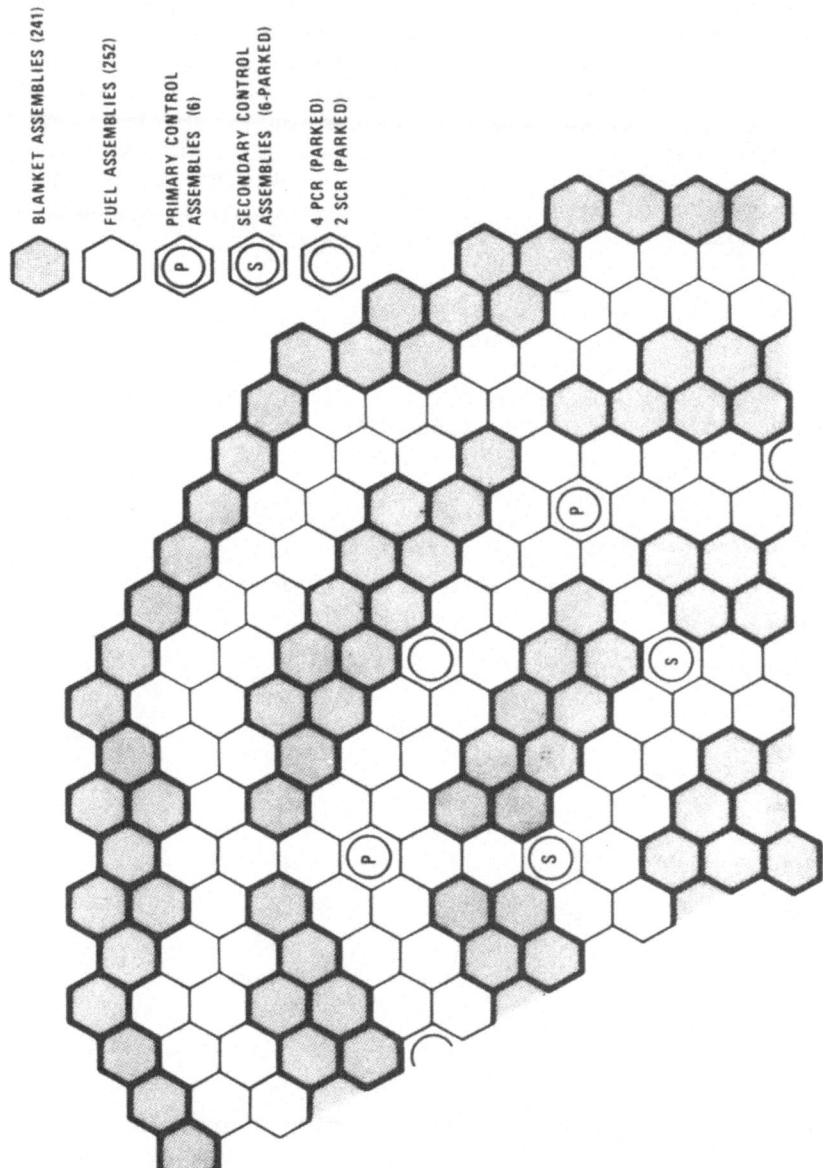

BLANKET ASSEMBLIES (241)

FUEL ASSEMBLIES (252)

PRIMARY CONTROL ASSEMBLIES (6)

SECONDARY CONTROL ASSEMBLIES (6-PARKED)

4 PCR (PARKED)
2 SCR (PARKED)

Figure 9. Design C, Heterogeneous Core Layout, 2550 MWt

Table IV compares the detailed physics parameters of each of these designs. Due to the arrangements of fuel and inner blanket assemblies in Designs A and C, only a single fuel enrichment was required to maintain a relatively flat radial power distribution. Design A has the lowest fissile mass and compound system doubling time. The primary advantage of Design B is its low fuel depletion reactivity. The corresponding principal advantage of Design C is its lower maximum fuel assembly sodium voiding reactivity worth relative to Designs A and B. All three designs have compound system doubling time os less than 20 years.

TABLE IV

COMPARISON OF DESIGNS, A, B AND C's
PERFORMANCE PARAMETER, 2550 MW(t) POWER RATINGS

Parameter		Design A	Design B	Design C
Number of Enrichment Zones		1	2	1
Fissile Mass (kg)		4586	4779	5221
Breeding Ratio		1.43	1.44	1.43
Compound System Doubling Time		16.0	16.0	18.0
Fuel Depletion Reactivity (745 FPD)		1.86%Δk	0.91%Δk	2.44%Δk
Peak Linear Power Generation Rate (kW/ft) with 3 sigma uncertainties plus overpower				
Fuel Rod	BOEC	17.2	18.6	17.4
	EOEC	15.9	14.4	16.3
Blanket Rod	BOEC	5.9	6.2	5.9
	EOEC	20.4	18.6	17.5
Total Control Assembly Worths		5.82	5.50	5.61
Maximum Sodium Void Reactivity in Fuel Assemblies ($)		3.2	∿3.2	∿2.0

The principal disadvantages of Design C are its excessively high fuel depletion reactivity, high fissile mass and longer doubling time. Design B is inferior to Design A in all performance characteristics except for a lower fuel depletion reactivity. Design A is thus the more nearly optimum arrangement with its lower fissile mass, good doubling time, single enrichment, and a fuel assembly sodium void worth approximately half that of a similar homogeneous configuration, although slightly larger than that of Design C.

The Design A configuration was analyzed further for its response to a hypothetical overpower transient with failure to scram. The potential for a highly energetic, hypothetical core disruptive accident is reduced for this radial parfait core arrangement because of both the reduced sodium void reactivity in the fuel assemblies and the temporal incoherence of voiding between the fuel and blanket assemblies. The relative response of the heterogeneous and homogeneous designs to a transient undercooled accident without scram was estimated with the transient analysis code FX-2(40) and an estimated voiding sequence. The results indicate that the hypothetical core disruptive accident (HCDA) energetics in the heterogeneous design are smaller by a factor of approximately seven compared with a similar homogeneous design. More detailed safety analyses using an SAS-3D(41) input to FX-2 would be expected to produce a greater difference between the energetics of the two designs.

VI. DESIGN FLEXIBILITY

In this section, the added design flexibility provided by the heterogeneous concept is presented. As will be demonstrated, significant flexibility is afforded the designer in terms of his ability to rearrange the fuel and inner blanket assemblies. The major limitation lies in the designer's inability to predict accurately the performance changes associated with small perturbations to a design. Two apparently minor changes in core arrangement that make significant changes to the overall performance of the reactor are shown, to illustrate the sensitivity of heterogeneous designs to small perturbations of the design. This extreme sensitivity is the major reason a designer's intuition is not always useful.

A. Increased Control Assembly Worths

A study was performed to determine if additional reactivity control could be achieved in the CRBRP heterogeneous

configuration by rearranging the same number of fuel and
inner blanket assemblies. The results of this study are
illustrated in Figure 10. The first heterogeneous configuration is the CRBRP design, presented in Section V. The
second configuration is a slight modification. The modification consists of interchanging 12 fuel and blanket assemblies between the last ring of inner blanket assemblies and
the next ring of fuel so more fuel assemblies are adjacent
to the primary control assemblies in the row-7 positions.
In addition, the six refueling locations adjacent to the six
secondary control assemblies at the row-7 corner positions
were moved to maintain the peak linear power in the blanket
assemblies below 20 kW/ft.

This minor rearrangement of fuel assemblies adjacent to
control assemblies results in a large increase in both the
primary and secondary control system worths. The reactivity
loss (% k) for a burnup of 550 (275 + 275) full power days
was reduced by approximately 10%, primarily due to a 0.03
increase in the breeding ratio. This increase in breeding
was primarily the result of the reduction in the fuel enrichment requirement and the concomitant increase in uranium-
238 in the core.

B.   Straight Burn Capability

As discussed in Section II, the CRBRP reference design
employs annual refueling. Annual refueling also was employed
in the first and second configurations noted in Figure 10.
A heterogeneous core configuration (third configuration in
Figure 10) employing a new fuel design would, insofar as
physics considerations are concerned, permit a straight 550
full power days of operation (two calendar years), elimin-
ating the partial refueling of six assemblies after 275 full
power days.

The following modifications would be required: (1), the
six refueling locations would be made permanent fuel assembly locations; (2), the central blanket assembly would be
made a fuel assembly; (3), three additional primary control
assemblies would be added to the row 4 positions; (4), the
fuel rod diameter would be increased to 0.240 inch; and (5),
the fissile loading would be increased. The net result of
these modifications would be a reactor having sufficient reactivity control to burn 550 full power days without refueling.

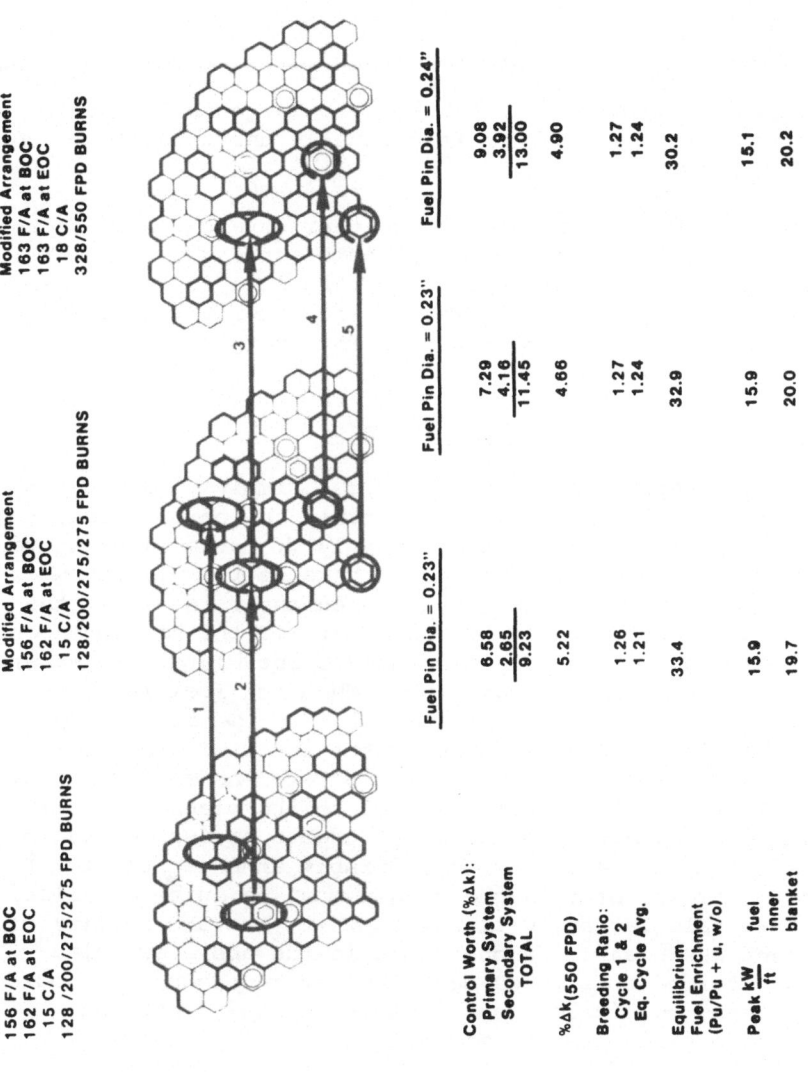

Figure 10. Influence of Heterogeneous Configuration on Core Performance

In addition, there would be a significant reduction in the fuel assembly peak linear power - from 15.9 to 15.1 kW/ft. The breeding ratio would be unchanged because the increased fertile uranium mass associated with the larger fuel pin diameter compensates for the increased fissile loading requirement.

The reactivity swing with burnup in the lower breeding ratio homogeneous design of CRBRP is so large that a 550 full power day straight burn is virtually unattainable, even with fuel redesign.

The primary advantage of a two-year straight burn capability in an LMFBR is the elimination of half the considerable time involved in preparing for refueling and reestablishing the operating configuration after refueling, independent of the actual time required for charging and discharging of the assemblies.

C.   Secondary Effects of Heterogeneity

If the heterogeneity is introduced simply by adding solid rings of blanket assemblies within the core region, the performance does not differ greatly from the alternative of increasing the fuel rod diameter because the primary mechanism by which the breeding ratio is enhanced is simply the addition of U-238 to the core. However, if the heterogeneous concept is used to provide azimuthal zoning, large increases in control worths can be achieved. When the fuel is clustered around the control rods, the flux in the cluster will be depressed early in life by the inserted control rod, avoiding excessive peaking in the cluster. As the burn proceeds, the flux shifts from the fuel to the blankets, so the removal of the control rod will not create excessive peaking factors. The net result is considerably enhanced control rod worths. Analysis of this phenomenon must be quite detailed; simple R-Z calculations will not show it; but the reality of the effect was clearly demonstrated in the above example. The two heterogeneous cases described above presumably would have the same "equivalent homogeneous reactor," but they themselves clearly are not equivalent.

In addition, the following factors make the two systems (heterogeneous and "equivalent" homogeneous) different. The lower flux in a heterogeneous reactor permits a decreased

interassembly gap for the same assembly wall thickness. This
additional space can be used for increased fuel rod diameter.
The magnitude of this advantage should decrease with time.
Designs utilizing the present stainless steel can make considerable use of the lower flux, but future alloy development
should reduce this advantage. On the other hand, the disadvantage of overcooling the inner blanket assemblies at beginning of life, to the detriment of fuel temperatures, can,
and probably will, be overcome through design innovations.

It is generally found that the peak-to-average burnup
is improved in a heterogeneous core. Since fuel life is limited by the peak burnup, this effect tends to reduce fuel
costs. In addition, a greater fraction of the total reactor
power is generated in the less-expensive blanket assemblies
in a heterogeneous design than in a homogeneous design. This,
too, tends to reduce fuel costs.

Offsetting these factors is the cost of additional blanket assemblies. Thus, cost effectiveness of each core depends upon relative fuel and blanket assembly costs, unless
a restriction is placed on allowable void reactivity. In
that case, the homogeneous core can be so distorted (pancake
core, for example) in order to meet the requirement that the
fuel cost can become excessively large.

VII.   HETEROGENEOUS EXPERIMENTAL VERIFICATION

ZPPR Assembly 7 (42,43,44,45) was the first critical
assembly in the United States (and probably the world) devoted primarily to the verification of fast reactor heterogeneous nuclear predictions. The experiments were performed
at the Argonne National Laboratory facility in Idaho Falls,
Idaho, from July 1976 through April 1978. This series of
experiments was conducted through a cooperative program of
the Argonne National Laboratory, the Westinghouse Electric
Corporation and the General Electric Corporation under the
direction of the Energy Research and Development Adminis-
tration (the Department of Energy since October 1977).

Each of the above participants analyzed the experiments,
and in general, obtained good agreement with the data. The
detailed analysis presented in this section was performed
at the Westinghouse Electric Corporation. Similar results

were obtained by the Argonne National Laboratory(46,47,48,49, 50,51) and the General Electric Corporation(51,52,53,54,55).

ZPPR-7 was designed to provide pre-engineering data of a heterogeneous design for a CRBRP-size reactor, with emphasis placed on those parameters sensitive to the heterogeneous core configuration. In order to achieve this purpose, the physical characteristics of the actual reactor configuration; i.e., region volumes, control locations and assembly pitch, and the hexagonal features of the CRBR were modeled as closely as possible in ZPPR-7 within the physical limits of the rectangular ZPPR matrix. The various phases of ZPPR-7 were:

Phase A (see Figure 11):   clean annular blanket
                           rings, no control rods

Phase B (see Figure 11):   beginning of life (BOL)
                           configurations with 12
                           control rod positions

Phase C (see Figure 12):   equilibrium core end-of-
                           life (EOL) conditions;
                           fuel drawers simulate
                           depleted plutonium and
                           inner blanket drawers
                           spiked to simulate plu-
                           tonium buildup, and 12
                           control rod positions

Phase D (see Figure 12):   same as Phase C except
                           number of control rods
                           increased to 15 and
                           blanket spiking modified.

Typical measurements performed were criticality, control rod worths, sodium voiding worths, and reaction rate distribution. Calculations were performed in both RZ and XY geometry using the two-dimensional diffusion theory code 2DB*(56). All calculations utilized ENDF/B-III nuclear data in either 9 or 21 neutron energy groups collapsed from 30-group data that were corrected for resonance self-shielding, elastic removal, and spatial flux weighting (cell heterogeneity). This data preparation sequence involves the use of the codes XSRES*(57), IDX*(57) and ANISN(58), respectively.

---

*See footnotes, p.84

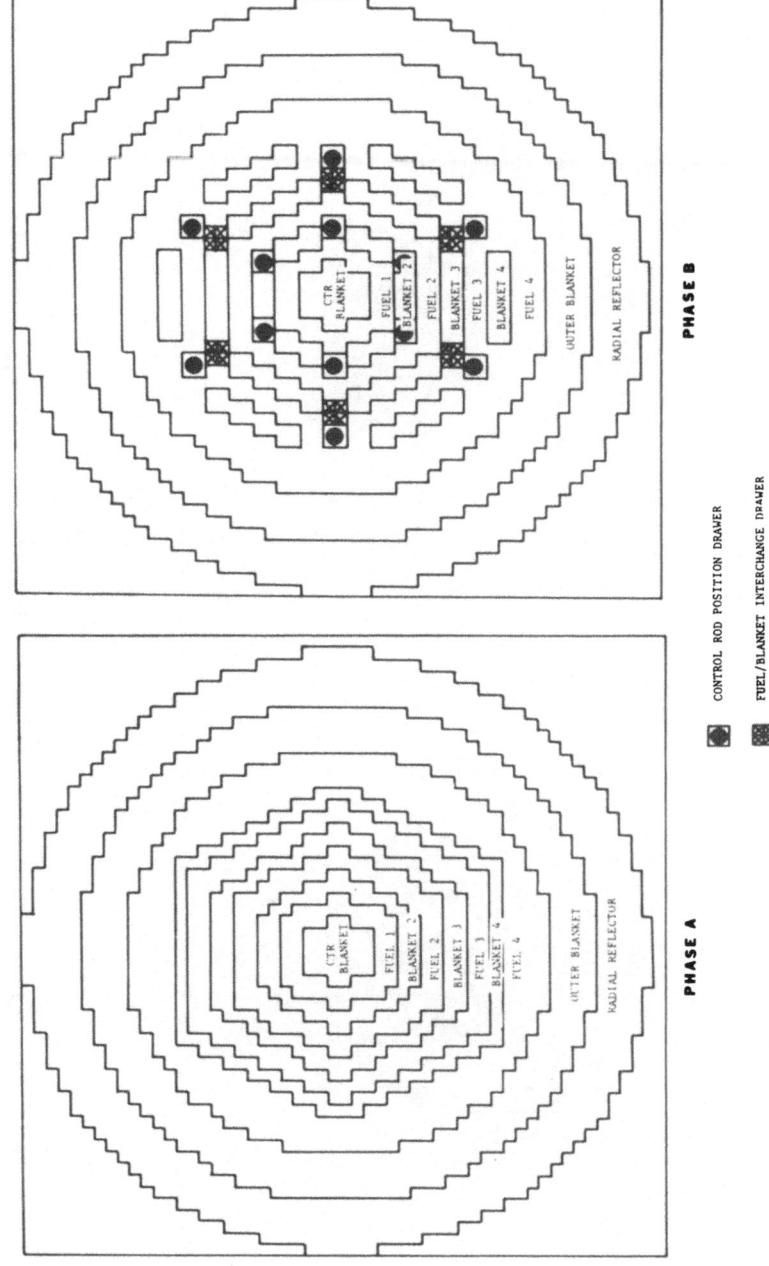

Figure 11. ZPPR-7 Critical Assembly Configurations, Phases A and B(42 and 43)

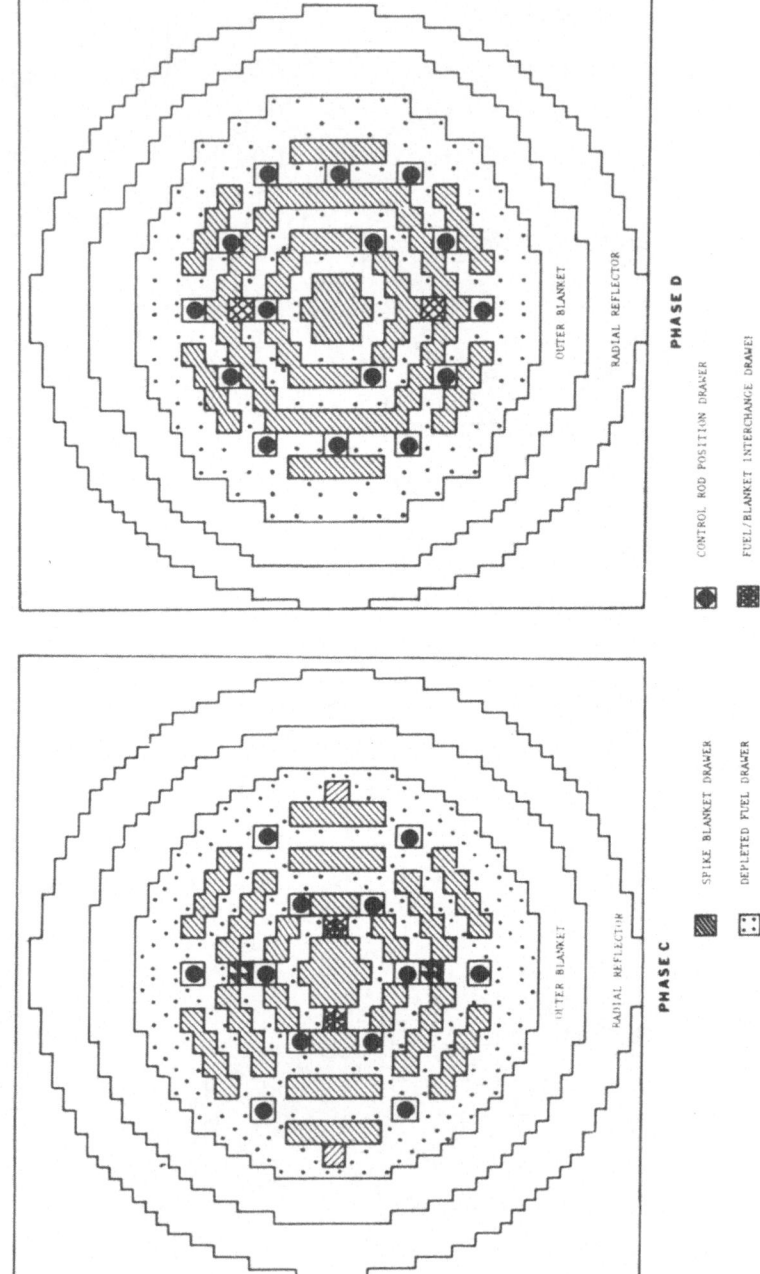

Figure 12. ZPPR-7 Critical Assembly Configurations, Phases C and D (44 and 45)

A comparison of the predicted versus experimental results is shown in Tables V through VIII. In general, the
results for the heterogeneous core show good agreement between predictions and experimental data. A similar study
was performed earlier in ZPPR-4, a mockup of the homogeneous
CRBRP. These values are compared in Table IX with the
heterogeneous values (Tables V through VIII) to gain insight
into the applicability of the analytical methodology to analyze either system. The calculated-to-experiment (C/E) ratios
vary somewhat for both concepts. The major difference is
that the heterogeneous C/E values for the control rods are
more spatially dependent than those for the homogeneous configuration, and approximately 10 percent lower in magnitude.
This type difference can be accommodated in a heterogeneous
design by proper arrangement of the fuel and inner blanket
assemblies relative to the control assemblies to assure sufficient control system worth.

In summary, no experimental results were found that
would inhibit the design of a heterogeneous liquid metal
fast breeder reactor or negate its higher breeding or lower
sodium voiding characteristics.

TABLE V

ZPPR-7 CRITICALITY RESULTS

Configuration	Geometry	K-Effective (Measured)	K-Effective (Calculated)	C/E
ZPPR-7A	R-Z	1.000278	0.98323	0.983
ZPPR-7A	X-Y	1.000278	0.99019	0.990
ZPPR-7B	R-Z	1.000636	0.97964	0.979
ZPPR-7B	X-Y	1.000636	0.98924	0.989
ZPPR-7C	R-Z	1.0000146	0.98112	0.981
ZPPR-7C	X-Y	1.0000146	0.99089	0.991
ZPPR-7D	X-Y	1.0000104	0.99347	0.993

TABLE VI

ZPPR-7 CONTROL ROD REACTIVITY WORTH

Control Rod Configuration	Phase	Measured Control System Worth ($)	Calculated Control System Worth ($)	C/E
6-Row 4 Inserted	B	14.00	12.77	0.912
6-Row 7C Inserted	B	17.90	17.62	0.984
6-Row 7F Inserted	B	13.40	12.74	0.891
6-Row 4 Inserted	C	22.12	19.97	0.903
6-Row 7C Inserted	C	14.35	12.93	0.901
6-Row 7F Inserted	C	21.22	20.03	0.944
3-Row 4 Inserted	D	11.00	9.75	0.886
6-Row 7C Inserted	D	14.35	12.99	0.905
6-Row 7F Inserted	D	15.41	13.78	0.894

Note: Phase B Values Simulate Beginning-of-Life (BOL) While Phases C and E are End-of-Life (EOL)

TABLE VII

ZPPR-7A SODIUM VOIDING WORTHS

Step	Region Voided	Axial Extent	Measured Worth (¢)	Calculated Worth (¢)	C/E
1	Fuel Ring 2	±12"	49.9	53.7	1.08
2	Fuel Ring 1	±12"	28.2	25.4	0.90
3	Blanket Ring 1	±12"	23.0	23.9	1.04
4	Fuel Ring 1	±12"-18"	-5.9	-4.2	----
5	Fuel Ring 1	±18"-24"	-7.5	-5.7	----
6	Central Blanket	±12"	9.3	7.8	0.84
7	Central Blanket	±12-18"	-0.5	----	----
8	Central Blanket	±18"-24"	-1.1	----	----

TABLE VIII

SUMMARY REACTION RATES FOR ALL ZPPR-7 PHASES

Reaction Rate	Region	Number of Date Values	Power Normalization Factor	Mean C/E	RMS Deviation
Pu-239 (n,f)	Core	106	1.000	1.000	± 1.76%
Pu-239 (n,f)	Inner Blankets	66	————	1.012	± 1.29%
U-235 (n,f)	Core	173	1.040	1.000	± 1.56%
U-235 (n,f)	Inner Blankets	93	-----	1.001	± 1.70%
U-238 (n,f)	Core	148	0.832	1.002	± 4.08%
U-238 (n,f)	Inner Blankets	92	-----	1.131	± 4.93%
U-238 (n,γ)	Core	148	1.097	1.002	± 2.56%
U-238 (n,γ)	Inner Blankets	92	-----	0.994	± 1.75%

TABLE IX

COMPARISON BETWEEN HOMOGENEOUS AND HETEROGENEOUS C/E VALUES

Parameter	Homogeneous	Heterogeneous
Criticality	0.9970[a]	0.9907[a]
Control Assembly Row 4 Corners	0.98[a]	0.91-0.89[a]
Control Assembly Row 7 Corners	1.00[a]	0.98-0.90[a]
Control Assembly Row 7 Flats	0.98[a]	0.89-0.92[a]
Reaction Rate 239 Pu (n,f)	+1.9%[b]	+1.8%[b]

Note: The values indicated for control assembly C/E ratios are for beginning-

of-life and end-of-life conditions respectively.

[a] Bias Factors are defined as E/C

[b] Uncertainty

VIII. APPLICABILITY TO ALTERNATE FUEL CYCLES

The breeder program in the United States, as in the
rest of the world, has been, to date, concentrated on the
Pu-239, U-238 cycle. This cycle offers two distinct advantages. First, the highest breeding ratio, lowest doubling
time, developed systems, are liquid metal fast neutron spectrum reactors fueled with plutonium and containing fertile
material of U-238 depleted of U-235 below the natural U-235
concentration.

Second, the plutonium-fueled LMFBR's are symbiotic with
light water reactors. The light water reactors require uranium enriched in U-235 for fuel. The enrichment process
naturally leaves large quantities of depleted U-238 available
for use as fertile material in LMFBR's. The fuel discharged
from an LWR contains a significant amount of Pu-239, the
fuel for LMFBR's, as well as significant quantities of unburned U-235. The Pu-239, and the unburned U-235, can be

concentrated and reused in the LWR's, but the Pu-239 is more
effectively utilized in an LMFBR.

The plutonium-fueled LMFBR system is not the only con-
ceivable nuclear power cycle. Many others are available,
but adequate treatment of the options is beyond the scope
of this article. A discussion of several alternatives is
available in reference 59. That same reference notes that
of all possible cycles the LWR system, without recycle, is
the most wasteful of uranium ore. With LMFBR's and recycle,
however, it provides both effective utilization of resources
and the potential for a fairly rapid growth of energy supply.

There is a concern about the use of plutonium, however.
The concern is that proliferation of nuclear weapon capability to non-nuclear weapon states will occur because of
access to plutonium in reactor fuel. While some hold the
opinion that plutonium systems have this risk, others claim
that the plutonium power reactor system is the least likely
road to proliferation, and that abandoning the plutonium
cycle in this country might add to the risk rather than minimize it. This subject is too lengthy to be treated in this
article. Rather, it is simply noted that alternate fuel
cycles that purport to minimize the risk of nuclear proliferation have been proposed as a replacement of the plutonium
cycle reactors. One well-known concept is that described by
Feiveson, Taylor et al(60).

Before a fuel cycle can be selected for development, detailed performance characteristics of the various reactors
used in it along with detailed mass flow balances must be
determined. The Department of Energy is presently engaging
several reactor manufacturers and national laboratories in
design studies aimed at determining the optimum overall cycle.
The final decision has not been made yet, but reasonable estimates of the roles of LMFBR's, and in particular, heterogeneous LMFBR's, can be inferred, based on the specific cycle
utilized.

One fuel cycle which proponents say effectively treats
the proliferation question and still permits reasonable energy growth is the so-called denatured cycle. In this cycle,
LWR's, both in the United States and abroad, are fueled with
either U-233 or U-235 as the fissile isotope, and use both
thorium and U-238 as the fertile isotopes. The amount of

U-238 used is five to six times the quantity of fissile uranium, rendering the material difficult for weapon use without further isotopic enrichment. Plutonium is formed as a byproduct of this cycle, but in quantities even smaller than that from the present LWR cycle. For example, a fuel mixture of 3.5% U-233, 20% U-238, and 76.5% Th would produce only about 20% as much Pu as would 3.5% U-235, 96.5% U-238 fuel. Presumably, the spent fuel from these reactors would be returned to the United States or other weapon states for reprocessing.

Once inside the United States or other weapon state, the spent fuel containing plutonium is shipped inside the boundary of an Energy Park. The Energy Park contains the reprocessor at which the plutonium is separated, a fabricator that manufactures plutonium-bearing fuel, and an LMFBR that is fueled with Pu-Th. This LMFBR burns the plutonium and breeds additional U-233 from the thorium. In this fashion the Energy Park serves as a transmuter or "plutonium sink", consuming plutonium bred outside the park. The assumption implicit in this rationale is that the Energy Park will be secure, and the potential for diversion from the park is substantially reduced.

The question of "diversion" is treated in a somewhat different manner. In this case, unacceptability for diversion could arise as a result of the contained radioactivity in the fabricated fuel. Spent fuel, for example, is more resistant to diversion because it is highly radioactive. Because U-232 if formed in a U-233-Th fueled reactor, the fuel refabricated of U-233-Th could be reasonably expected to emit radiation to the extent that it would be an unlikely candidate for diversion. Thus, it would be reasonable to locate a U-233-Th fueled LMFBR outside the Energy Park but within the boundaries of the United States to avoid proliferation. If desired, it would be possible to denature the fuel to render diversion and successful weapon production even more difficult.

These alternate fuel cycles were evaluated using the PLBR heterogeneous design identified in Figure 7. The different fuel cycles investigated are listed in Table X. Both the oxide and carbide form of the fuel material were examined. Concepts 1 and 2 utilize plutonium as the fissile material and uranium-238 as fertile material. This reference fuel

cycle results in the largest breeding gain for both the
oxide and carbide fuels. The replacement of the fertile
uranium-238 with thorium (Cases 3 and 4) results in reduc-
tions in the breeding gains. However, these reactors burn
large quantities of plutonium (623 and 651 kg/year) while
producing large amounts of fissile uranium (835 and 951
kg/year). Thus, this cycle is very attractive for the LMFBR
utilized within the Energy Park.

The pure uranium/thorium cycle (Concepts 5 and 6) results in a further reduction in breeding gain. It should
be noted, however, that the heterogeneous concept has not
been reoptimized for the uranium/thorium cycle. A breeding
ratio of 1.2 should be achievable by the use of a larger number of radial blanket assemblies and a thicker axial blanket.
Concepts 7 and 8 employ the denatured cycles that would be
used outside the Energy Park. They burn 150 to 239 kg/year
of uranium-233, while breeding 343 to 441 kg of plutonium
per year. A comparison of the mass gains of Concepts 3 and
4 indicates that one heterogeneous reactor inside the Energy
Park (Concept 3) can supply sufficient uranium-233 for approximately 1.9 denatured heterogeneous reactors (Concept 8)
outside the Energy Park, with sufficient excess plutonium
and uranium-233 for 3.9% per year growth rate of the symbiotic
reactor system. The overall system growth rate is reduced
to 2.9% per year (CSDT = 24) by inclusion of the out-of-
pile inventories.

The doubling time of the heterogeneous LMFBR, using the
plutonium/uranium oxide system (Concept 1) is of the order
of 16 years. The similar doubling time for pure uranium or
thorium cycle (Concept 5) is approximately 50 years. These
two doubling time extremes give some insight on the overall
detrimental impact that this proliferation-resistant fuel
cycle concept will have on electrical growth potential. Note
that the concept of "doubling time" lacks its usual meaning
in these cases. If the fuel is plutonium and the fissile
species generated is uranium-233, to be used in a different
reactor design, the doubling time is not the same length of
time as if plutonium were being generated at the same rate.
Nevertheless, the quantity provides insight to performance.

TABLE X

PERFORMANCE OF ALTERNATE FUELS
IN A 2550 MW(t) HETEROGENEOUS REACTOR CONCEPT

	1	2	3	4	5	6	7	8
Fuel Assembly	(Pu-U)O$_2$	(Pu-U)C	(Pu-Th)O$_2$	(Pu-Th)C	(U-Th)O$_2$	(U-Th)C	(U-233, U-238)O$_2$ Denatured	(U-233, U-238)O$_2$ Denatured
Inner and Radial Blankets	UO$_2$	UC	ThO$_2$	ThC	ThO$_2$	ThC	ThO$_2$	ThO$_2$ Denatured
Axial Blankets	UO$_2$	UC	ThO$_2$	ThC	ThO$_2$	ThC	ThO$_2$	ThO$_2$ Denatured
Core Height (ft)	4	3	4	3	4	3	4	4
Refueling Interval (Years)	2.5	3	3	3	Annual	Annual	Annual	Annual
BOEC (Fissile Mass/kg)	4525	3875	5264	4660	5136	4666	4488	4444
Average Breeding Ratio	1.44	1.62	1.32	1.44	1.16	1.19	1.25	1.26
Plutonium Gain/Year (kg)	310	427	-623	-651	----	----	343	441
Uranium Gain/Year (kg)	----	----	835	951	125	147	-150	-239
Net Fissile Gain/Year (kg)	310	427	212	300	125	147	193	202

The designs presented here are all heterogeneous designs. It appears almost certain that the Pu-Th transmuter will be a heterogeneous design to maximize uranium-233 production. However, the doubling time of the (U-233, Th)O_2 system is so long that improvement in doubling time or even breeding ratio by use of a heterogeneous design will not materially affect growth rate potential for many years. Rather, considering the very limited amount of U-233 available, growth will be enhanced through use of minimum inventory design.

IX. FUTURE OF HETEROGENEOUS CORES

To evaluate the future of heterogeneous cores, as compared with homogeneous cores, a quantitative figure of merit should be used to compare reactor designs. The usual parameters of comparison are economics, breeding ratio, specific inventory, doubling time and sodium void reactivity. In this section, these parameters are discussed. The authors' personal opinion of the relative merits of the homogeneous and heterogeneous concepts and the future of heterogeneous cores in the United States also is presented.

A.   Figures of Merit

1. Economics. According to the free market theory, the availability or scarcity of fuel should change its value so that the market would dictate the worth of inventory and breeding ratio, the two principal ingredients of doubling time. Hence, economics should be the principal figure of merit. However, the free market is not operable in the plutonium business, and if it were, it would probably produce wild swings in value. In addition, the cost of plutonium and its resale value is a relatively small portion of fuel cost, while fuel costs themselves are a relatively small portion of the total power cost.

Plutonium price does not depend on supply and demand for two reasons: one is that plutonium is under a government monopoly; a second is that the time lag between plutonium production in current light water reactors and a potential plutonium shortage is so long as to distort prices. It is quite likely that early commercial LMFBR's will have requirements set for minimum breeding and doubling time performance, even if not cost effective, to assure a future supply of energy for the good of the nation. Alternately, for the same reason, the government could elect an artificially higher buy-back price of plutonium during the early phase of the industry.

Nevertheless, economic optimization codes are used to produce "cost-optimized" designs. Economic considerations certainly are important; the lower cost design of two with greatly different cost projections clearly is preferred, but the determination of an absolute minimum cost core design is sufficiently inaccurate, even if the correct plutonium cost were assumed, that small differences in calculated cost cannot be regarded as significant.

Power cost estimates for homogeneous and heterogeneous core designs do differ. However, the observed differences are small, dependent on the specific heterogeneous design, and do not invariably favor either concept; thus, economics does not provide guidance for the selection of either concept. The other figures of merit must be considered.

2. <u>Breeding ratio</u>. The definition of breeding ratio (61) used in this article is the ratio of the production of fissile plutonium (Pu-239 + Pu-241) to the destruction of fissile material (U-235 + Pu-239 + Pu-241). In this definition, Pu-240, Pu-242 and the mass of U-235 up to the weight fraction found in depleted uranium (0.2%) are not considered as fissile material.

With this definition, an unambiguous quantitative value is obtained. It depends primarily on the cross-section set used. The current version of the Evaluated Nuclear Data File is the accepted set. Any difference codes of calculation or methods can be and have been resolved with benchwork calculations. As has been demonstrated in this article, the breeding ratio for a heterogeneous design, employing the same fuel assembly design, is approximately 0.1 higher than that of a homogeneous reactor of the same power level. This is one of the advantages of the heterogeneous concept.

3. <u>Plutonium inventory</u>. As indicated in Section III, the most detrimental characteristic of a heterogeneous core is the greater (∿25%) fissile inventory required. This is primarily the result of the higher effective fertile uranium concentration in the reactor (fuel and inner blanket assemblies).

However, if plutonium becomes limited, it is possible to
modify a heterogeneous design to have a lower inventory
while still maintaining a low fuel sodium void reactivity
and the incoherence of boiling. A suggested approach(62)
to reduce the impact of the higher fissile requirement would
be to reduce the fuel rod diameter from 0.31 inch to 0.23
inch, for example, and employ more rods (397 versus 271) in
each assembly. In this example, the number and arrangement
of fuel and blanket assemblies were maintained, as was the
total power. Thus, the average linear power in the fuel
rods was reduced. This reduction was utilized to reduce the
core height from four feet to three feet. This change reduced the reactor plus out-of-pile inventory by approximately 15%. In this analysis, the 0.23-inch fuel rod was
assumed to have a two-year lifetime, while the lifetime of
the 0.31-inch assembly was three years. Because of the larger number of rods and the reduced fuel lifetime, there was
a large increase in fuel cycle cost (approximately 50%).
This penalty may be reduced by the introduction of advanced
alloys with increased fuel burnup capability. The breeding
ratio was reduced from 1.42 to 1.32 and the doubling time
was increased from 17 to 19 years. In addition, the shorter
core height (four to three feet) further reduced the sodium
void reactivity from $3 to $2.

The French are taking a similar approach by design-
ing with their well-proven 0.23-inch diameter rod. Addition-
ally, they are employing a central axial blanket to improve
their power peaking and minimize fuel inventory. Their
overall conclusion on inventory is that they can achieve
heterogeneous inventories very close to that of a similar
homogeneous reactor. It should be noted, however, that
present homogeneous reactors are designed primarily to minimize doubling time, not inventory.

4. Doubling time. The problem with breeding ratio is
that Design A might breed better than Design B, but requires
so much more inventory that Design B has a lower doubling
time. Consequently, doubling time has become a more favored
measure of design worth than any other figures of merit.
Doubling time is the time for a reactor to produce sufficient
additional fuel, after losses, to refuel another identical
reactor, including pipeline requirements.

Doubling time would be the ideal figure of merit if

all of the available plutonium were used in the first generation LMFBR's and additional LMFBR's of the same design were
built as fast as plutonium became available to fuel them.
Doubling time would be thus a measure of the growth of LMFBR
power, with additional power for the nation produced by less
desirable, more costly systems. Unfortunately, this ideal
situation is the only one for which doubling time is a precise figure of merit.

If two reactors each have doubling times shorter
than the doubling of LMFBR power requirements, there is no
advantage to the reactor that has the shorter doubling time,
assuming no plutonium is being fed to other reactors. In
this latter case, breeding ratio comes back into the picture
as will be shown later. Similarly, if both have doubling
times so long that most growth can come primarily only from
LWR-produced plutonium, very little additional merit can be
ascribed to the one with the shorter doubling time.

Finally, the meaning of doubling time is obscure,
because the quantity that must be doubled is not plutonium,
but electrical power. To clarify this point, consider a
hypothetical case wherein the higher inventory core has a
slightly greater doubling time; that is, reactor design A
has a Pu inventory requirement of 7,000 kg and a 20-year
doubling time, and design B, of the same power, has a Pu
inventory requirement of 5,000 kg and an 18-year doubling
time. The doubling time figure of merit suggests design B
is superior. Assume there is enough plutonium available to
build a given required power capacity with design A reactors,
all built instantaneously. Assume further that design B
reactors are used for all power growth requirements. That
choice would lead to a 126% increase in power in 18 years.
Conversion of all design A cores to design B cores would add
another 40% for a total power increase of 166%. If design B
reactors were chosen for the first power production capability, the increase in power would be only 140% in the same
18 years. The first 100% comes from doubling the inventory
used and the extra 40% is the plutonium inventory not needed
in the first group of design B reactors to produce the same
initially required power capacity. Obviously, the shorter
doubling time system fails in this example because all of the
plutonium available for the first generation was not used to
build more initial reactors, on the assumption the initial
demand did not require the extra plants. Thus, the <u>power</u>

doubling time is actually lower in the scenario of starting
with design A reactors and adding design B reactors than if
design B reactors were chosen for the first generation.

Clearly, if the doubling times are the same for
both reactors, the higher inventory system will produce an
even greater increase in power capacity in the future,
assuming there is enough plutonium for the first generation
reactors. The conclusion is that if a plutonium shortage
is projected for some point in time beyond the expected life
of a reactor, it is worth the inventory penalty to improve
breeding ratio, independent of doubling time. Thus, doubling time is a precise figure of merit only for the idealized situation described earlier; while useful, it cannot
be considered an absolute measure of worth of a design without evaluation of the power growth scenario, and as discussed later, consideration of the mix of reactor systems.

In this article, the doubling time reported for a
large heterogeneous design is approximately two years lower
than that of a similar homogeneous reactor. Other publications have shown results that indicate either the same or
longer doubling times for a heterogeneous concept. The results are strongly dependent on the design of the heterogeneous concept and degree of decoupling employed (see
Section IV). In either case, the range of differences in
published doubling times between homogeneous and heterogeneous concepts is so small that it cannot be useful in the
selection of either concept.

5. Sodium void reactivity. The heterogeneous concept
has been shown to reduce the fuel sodium void reactivity by
approximately half, relative to a homogeneous reactor. In
addition, the temporal incoherence of voiding between the
fuel and blanket assemblies results in a considerable reduction in the initiating reactivity available during an HCDA.
This probably is the most significant positive merit of a
heterogeneous reactor. However, it is very difficult to
evaluate as a figure of merit since it is believed that homogeneous designs are quite adequate from the standpoint of
HCDA probability and magnitude of energetics. However,
according to analyses performed to date, even more conser-
vative assumptions must be made to calculate energetic results
from an HCDA in a heterogeneous reactor than from an HCDA in
a homogeneous reactor. Thus, reduced sodium void reactivity

might have a positive effect on the licensing process of
commercial LMFBR's.

B.   Future of the Heterogeneous Concept

The choice of a figure of merit is totally dependent
upon the scenario assumed. This can be illustrated best by
assuming a hypothetical case for the purpose of illustration
where there is no electrical power growth; new reactors are
replacements requiring no new inventory, all stationary power
plants are breeders, but propulsion reactors are nonbreeders.
If the ratio of nonbreeder to breeders is to be maximized,
breeding ratio would be the only figure of merit (other than
cost). This is a purely hypothetical case of course, since
a cost-optimized system would include a mix of LMFBR's and
LWR's. Nevertheless, it does show the degree to which the
power scenario affects the selection of the best figure of
merit.

In another case, if an acute plutonium shortage existed,
compared with the amount required for new power plants, inventory would be the overriding figure of merit. This is
not a purely hypothetical case. For example, England is
moving ahead fairly rapidly with commercialization of LMFBR's.
They expect plutonium to be limited so that inventory is
likely to be the most important parameter in UK reactors.
One would also expect the French to be plutonium-limited,
due to their aggressive commercialization. In the United
States, however, inventory may not be very important, particularly for the first few reactors, depending on the growth
rate of LMFBR's (assuming sufficient reprocessing capability
is developed).

The other figures of merit, cost and doubling time, were
shown to be little different between heterogeneous and homogeneous designs as long as there is no requirement on sodium
void reactivity. If there is a sodium void reactivity restriction, the choice clearly is the heterogeneous design.

The conclusion to be drawn from evaluating the quantitative parameters is that no clear choice can be made between
heterogeneous and homogeneous cores with the present most
likely scenarios for LMFBR growth in this country.

Nevertheless, it is very likely that early commercial

LMFBR's built in this country will be heterogeneous. As
noted earlier, heterogeneous design performance is relatively independent of fuel rod size. This can be a significant
licensing advantage, because a change in fuel design may
require verification testing that could cause delays. The
potential for an energetic hypothetical core disruptive
accident is vanishingly small for either reactor design.
However, the potential for an energetic hypothetical core
disruptive accident is reduced significantly for a heterogeneous design; thus, it no doubt will be easier and thus
faster, to license a plant with a low sodium void reactivity
and even lower calculated energetics as a result of a transient overpower with failure to scram. To achieve the same
results in a homogeneous design requires an extremely off-
optimum system (pancake reactor) that is costly and performs
poorly.

The licensing advantage of heterogeneous cores can be
converted directly to a cost advantage. Capital investment
costs are the largest fraction of LMFBR power costs. Of the
capital cost, the largest single item is interest during
construction; hence, construction delays increase power cost
significantly. The lead time for a nuclear reactor is paced
primarily by licensing; thus, any design feature that might
expedite the licensing process is indeed valuable, and the
heterogeneous concept is one such important design feature.

FOOTNOTES

*   Section III. The Atomic Energy Commission (AEC) was
     organized in 1947. It became a part of the Energy
     Research and Development Administration (ERDA) on
     January 19, 1975. More recently, October 1, 1977, it
     became part of the Department of Energy (DOE). All
     three names are used here, depending upon the date
     of the action. For the efforts discussed in this
     article, the three are synonomous.

*   Section VII. The codes used are Westinghouse
     proprietary modifications of BNWL Codes. The
     Westinghouse modifications are proprietary and
     are identified as an 'excepted item' under DOE
     contacts with Westinghouse.

ACKNOWLEDGEMENTS

The authors thank the members of the Nuclear Design Group at the Westinghouse Advanced Reactors Division for their contributions to this paper. The Westinghouse analyses reported in the paper were performed by V. Arakali, G. J. Calamai, J. A. Lake, D. Lancaster, N. C. Paik, D. S. Petras, R. W. Rathbun, R. Rittenberger, R. W. Rupnik, S. K. Varner and M. Yarbrough in their respective areas of expertise. The authors also express appreciation to D. Felix, M. Frazee and J. Proietti for their editorial and graphical contributions.

REFERENCES

1. Ducat, G. A., Driscoll, M. J. and Todreas, N. E., "Evaluation of the Parfait Blanket Concept for Fast Breeder Reactors," COO-2250-5, January, 1974.

2. Driscoll, M. J., Ducat, G. A., Pinnock, R. A. and Aldrich, D. C., "Safety and Breeding-related Aspects of Fast Reactor Cores having Internal Blankets," in Fast Reactor Safety and Related Physics, Vol. II, CONF-761001, PP 575-583.

3. Naser, J. A., Sehgal, B. R., Winkleblack, K., "LMFBR Heterogeneous Core Designs with Axial Internal Blankets," Trans. American Nuclear Society, 26, PP 561-562, 1977.

4. Estavoyer, M., Ravier, J., Chaumont, J. M., Mougniot, J. C., Marmonier, P. P., Sicard, B., Gourdon, J., Delpeyroux, P., Savineau, M., Clauzon, P., "Evolution of French Fast Neutron Reactor Core Design and Performance," International Conference on Optimization of Sodium-cooled Fast Reactors, PP 145-151, British Nuclear Energy Society, London, 1977.

5. Novick, M., McGinnis, F. D., Whitham, G. K., "EBR-I and EBR-II Operating Experience," Fast Reactor Technology, pp 25-40, American Nuclear Society, Hinsdale, Illinois, 1965.

6. Meneghetti, D. and Loewenstein, W. B., "EBR-II Physics Experience," International Symposium on Physics of Fast

Reactors, Vol. 1, pp 175-199, Committee for International
Symposium on Physics of Fast Reactors, Tokyo, Japan,
1973.

7.   "Retirement of the Enrico Fermi Atomic Power Plant,"
      NP-20047, March, 1974 (availability: U. S. DOE Tech-
      nical Information Center).

8.   Alexanderson, E. L., Branyan, C. E., Olson, W. R.,
      "Operating Experience at the Enrico Fermi Atomic Power
      Plant," Fast Reactor Technology, pp 41-46, American
      Nuclear Society, Hinsdale, Illinois, 1965.

9.   Caldarola, L., Freeman, D. D., Greebler, P., Kussmaul,
      G., Mitzel, F., Noble, L. D., Pflasterer, G. R.,
      Oosterkamp, W. J., Wintzer, D., "SEFOR Experimental
      Results and Application of LMFBR's," in Engineering of
      Fast reactors for Safe and Reliable Operation, Vol. III,
      PP 1312-1330, Gesillschaft fur Kernforschung, MBH,
      Karlsruhe, West Germany, 1973.

10.   Arterburn, J. O., Hikido, K., "SEFOR Operating Exper-
      ience and Significance to LMFBR's," Engineering of Fast
      Reactors for Safe and Reliable Operation, Vol. II, PP
      510-521, Gesillschaft fur Kernsforschung, MBH, Karlsruhe,
      West Germany, 1973.

11.   "FFTF Final Safety Analysis Report," Richland, Washing-
      ton, HEDL-TI-75001, 1, Figure 2.1-2, December, 1975.
      (Availability: U. S. DOE Technical Information Center).

12.   Clinch River Breeder Reactor Project, Preliminary Safety
      Analysis Report, Docket-50537 1, 1975.

13.   Krasik, S. and Radkowsky, A., "Pressurized Water Reactor
      (PWR) Critical Experiments," Proceedings of the inter-
      national Conference on the Peaceful Uses of Atomic Ener-
      gy, Geneva, 1955, Vol. 5, pp 203-214, United Nations, New
      York, 1955.

14.   Radkowsky, A. and Krasik, "Physics Aspects of the Press-
      urized Water Reactor (PWR)," Proceedings of the Inter-
      national Conference on the Peaceful Uses of Atomic Ener-
      gy, Geneva, 1955, 5, pp 229-238, United Nations, New
      York, 1955.

15. Radkowsky, A. and Bayard, R. T., "The Physics Aspects
     of Seed and Blanket Cores with Examples from PWR,"
     Proceedings of the Second United Nation's International
     Conference on the Peaceful Uses of Atomic Energy,
     Geneva, 1958, Vol. 13, pp 128-145, United Nations, New
     York.

16. Radkowsky, A., Hardigg, G. W. and Luce, R. G., "Seed
     and Blanket Reactors," Proceedings of the Third Inter-
     national Conference on the Peaceful Uses of Atomic
     Energy, Vol. 6, pp 304-311, Geneva, 1964, United Nations,
     New York, 1965.

17. "Liquid Metal Fast Breeder Reactor Design Study,"
     WCAP 3251-1, 1964.

18. "Large Fast Reactor Design Study," ACNP-64503, 1964.

19.   Liquid Metal Fast Breeder Reactor Design Study (1000
     MWe UO -PuO Fueled Plant)," GEAP-4418, Vol. I and II,
     1964.

20. Liquid Metal Fast Breeder Reactor Design Study, CEND-200,
     Vol. I and II, 1964.

21. "Conceptual Plant Design, System Descriptions and Costs
     for a 1000 MWE Sodium-cooled Fast Reactor, Task II,
     Report of 1000 MWe LMFBR Follow-on Work," GEAP-5678,
     December, 1968.

22. "1000 MWE LMFBR Design Modification for Oxide Core,"
     Westinghouse Advanced Reactors Division, Madison, Penn-
     sylvania, WARD-127, July, 1970.

23. Sehgal, B. R., Lin, C., Loewenstein, W. B., "A High
     Breeding Ratio Low Sodium Void Coefficient LMFBR Core,"
     Trans. American Nuclear Society, 23, pp 401-402, 1976.

24. Moucniot, J. C., Barre, J. Y., Clauzon, P. Ciacometti,
     C., Neviere, G., Rovier, J., Sichard, B., "Breeding
     Gains of Sodium-cooled Oxide-fueled Fast Reactors,"
     presented at the European Nuclear Conference, Paris,
     April, 21-25, 1975 (ORNL-tr-2994).

25.   Doncals, R. A., Paik, N. C., Dickson, P. W., "Trans-
      position of Fuel and Blanket Assemblies in LMFBR's,"
      presented at the American Nuclear Society Annual
      Meeting, New York, New York, June 12-17, 1977.

26.   Dickson, P. W., "Engineering Aspects of Heterogeneous
      Reactors," International Conference on Optimization of
      Sodium Cooled Fast Reactors, pp 207-216, British
      Nuclear Energy Society, London, 1977.

27.   Henninger, R. J., Bowers, C. H., Cahalan, J. E.,
      Ganapol, B. D., Wang, W. L., Ferguson, D. R., "An
      Analysis of Unprotected Loss-of-Flow Accidents for the
      CRBRP with a Radial Parfait Core," Trans. American
      Nuclear Society, 24, pp 260-261, 1976.

28.   Tzoros, Constantine P., Barthold, W. P., "Sensitivity
      of the Power Distribution in Large Heterogeneous LMFBR
      Designs," Trans. American Nuclear Society, 26, pp 544-
      545, 1977.

29.   Tzoros, Constantine P., Barthold, W. P., "A Systematic
      Approach for Constructing Low Sodium Void Heterogeneous
      Cores," Trans. American Nuclear Society, 26, pp 548-
      549, 1977.

30.   Barthold, W. P., Beitel, J., Khan, E., Tzanos, C.,
      "Potential and Limitation of the Heterogeneous Reactor
      Concept," Trans. American Nuclear Society, 26, pp 552-
      553, 1977.

31.   Bailey, H. S., Lu, Y. S., "Nuclear Performance of LMFBR's
      Designed to Preclude Energetic HCDA's," Trans. American
      Nuclear Society, 26, June, 1977.

32.   Vitti, J. A., Felten, L. D., Galluzzo, N. G., Otter, J.
      W., "Nuclear Design and Economic Comparison of a Con-
      ventional and Bullseye LMFBR Core," Trans. American
      Nuclear Society, 26, pp 553-554, 1977.

33.   Paxson, E., "Radial Parfait Core Design Study," Westing-
      house Advanced Reactors Division, Madison, Pennsylvania,
      WARD-353, June, 1977.

34. Sicard, B., Mougniot, J. C., Sztark, H., Cabrillat,
     J. C., Giacometti, C., Carnoy, M., Clauzon, P.,
     "Preliminary Physics Studies of a Large Fast Core
     Based on the Heterogeneous Concept," Trans. American
     Nuclear Society, 26, Page 553, 1977.

35. Spenke, H., Wehmann, U., Pilate, S., de Wouters, R.,
     Kiefhaber, E., "Physics Studies of a Heterogeneous
     Core Concept for SNR-2," Trans. American Nuclear
     Society, 26, Page 561, 1977.

36. Bouget, Y. H., Desprets, A., Hammer, P., Humbert, G.,
     Lyon, F., Wiesenfeld, B., Martini, M., Cosimi, M.,
     Brocoli, U., "Physics Performance of Heterogeneous
     Fast Reactor Core Concept Studied in MASURCA," Trans.
     American Nuclear Society, 26,   554-555, 1977.

37. Bruna, G. B., Cecchini, G. P., Gastaldo, G., Kallfelz,
     J. M., Palmiotti, G., Salvatores, M., Rowland, D.,
     Bartine, D., Burns, T., "Studies of the Heterogeneous
     LMFBR Core Concept for Several Fuel Cycles and Its
     Sensitivity to Design and Cross-Section Changes,"
     Trans. American Nuclear Society, 26,   558-559, 1977.

38. Calamai, G. J., Varner, S. K., Doncals, R. A., "Optim-
     ization of Fuel Resources in Commercial Oxide LMFBR's,"
     Trans. American Nuclear Society, 23,   434-435, 1976.

39. Barthold, W. P., Tzanos, C. P., Beitel, J. C., "Poten-
     tial of Large Heterogeneous Reactors," International
     Conference on Optimization of Sodium-cooled Fast Reactors,"
     pp 217-227, British Nuclear Energy Society, London, 1977.

40. Meneley, D. A., Leaf, G. K., Lindeman, A. J., Daly, T.
     A., and Sha, W. T., "A Kinetics Model for Fast-Reactor
     Analysis in Two Dimensions," Dynamics of Nuclear Systems,
     pp 483-500, The University of Arizona Press, Tucson,
     Arizona, 1972.

41. Dunn, F. E., Fischer, G. J., Heames, T. J., Pizzica, P.,
     McNeal, N. A., Bohl, W. R. and Prastein, S. M., "The
     SAS-3A LMFBR Accident Analysis Computer Code," AN/RAS
     75-17, April, 1975.

42. "Reactor Development Program Progress Report, July,
     August, 1976," ANL-RDP-52, October, 1976. (Availa-
     bility: U. S. DOE Technical Information Center).

43. "Reactor Development Progress Report, September, 1976,"
     ANL-RDP-53, November, 1976. (Availability: U. S. DOE
     Technical Information Center).

44. "Reactor Development Program Progress Report, October,
     1976," ANL-RDP-54, December, 1976. (Availability:
     U. S. DOE Technical Information Center).

45. "Reactor Development Program Progress Report, November,
     1976, ANL-RDP-55, January, 1977. (Availability: U. S.
     DOE Technical Information Center).

46. "Reactor Development Program Progress Report, 1976,
     ANL-RDP-56, February, 1977.

47. "Reactor Development Program Progress Report, January,
     1977," ANL-RDP-57, March, 1977. (Availability: U. S.
     DOE Technical Information Center).

48. "Reactor Development Program Progress Report, February,
     1977," ANL-RDP-58, April, 1977. (Availability: U. S.
     DOE Technical Information Center).

49. "Reactor Development Program Progress Report, March,
     1977," ANL-RDP-59, May, 1977. (Availability: U. S.
     DOE Technical Information Center).

50. "Reactor Development Program Progress Report, July,
     1977," ANL-RDP-60, April, 1977. (Availability: U. S.
     DOE Technical Information Center).

51. Lineberry, M. J., Carpenter, S. G., McFarlane, H. F.,
     Collins, P. J., Amundson, P. I., Stewart, S. L.,
     Hitchcock, J. T., "ZPPR Critical Experiments for a
     Radial Parfait Reactor," Trans. American Nuclear Society
     26, pp 555-556, 1977.

52. "Critical Experiments and Analysis Twenty-first Quarter-
     ly Report October-December 1976," GEAP-13771-21, January
     1977. (Availability: U. S. DOE Technical Information
     Center).

53. "Critical Experiments and Analysis Twenty-second
    Quarterly Report January-March 1977," GEFR-13771-22,
    April, 1977.

54. "Critical Experiments and Analysis Twenty-third Quar-
    terly Report April-June 1977," GEFR-13771-23, July,
    1977. (Availability: U. S. DOE Technical Information
    Center).

55. Hartman, A. K., Hitchcock, J. T., Stewart, S. L.,
    "Control Rod Worths in a Simulated Radial Parfait
    LMFBR Core," Trans. American Nuclear Society, 26,
    pp 538-539, 1977.

56. Little, W. W., Jr., Hardie, R. W., "2DB User's Manual,
    Revision 1," BNWL-831, August, 1969.

57. Hardie, R. W., Little, W. W., Jr., "1DX: A One-
    dimensional Diffusion Code for Generating Effective
    Nuclear Cross-sections," BNWL-954, March, 1969.

58. Engle, W. W., Jr., "A User's Manual for ANISN: A One-
    dimensional Discrete Ordinates Transport Code with
    Anisotropic Scattering," K-1693, March, 1967.

59. Metz, W. D., "Reprocessing Alternatives: The Options
    Multiply," Science, 196, pp 284-287, April, 1977.

60. Feiveson, H. A., Taylor, T. B., Hippel, F., Williams,
    R. H., "The Plutonium Economy: Why We Should Wait and
    Why We Can Wait," The Bulletin of the Atomic Scientists,
    32, No. 10, pp 10-13, 1976.

61. Lake, J. A., Doncals, R. A., Rathbun, R. W., Robinson,
    H. C., "Breeding Ratio and Doubling Time Characteristics
    of the Clinch River Breeder Reactor," Advanced Reactors:
    Physics, Design and Economics, pp 665-676, Pergamon
    Press, New York, 1975.

62. Paik, N. C., Davis, W. J. and Bortz, A. B., "Physics
    Evaluations and Applications, Quarterly Progress Report
    for Period ending January 31, 1977," WARD-XS-3045-16,
    1977. (Availability: U. S. DOE Technical Information
    Center).

LINER INSULATION FOR GAS-

COOLED REACTORS

B. N. Furber and J. Davidson

Nuclear Power Company (Risley) Limited

Risley, Warrington, England

I.   INTRODUCTION

A.   Advantages of Concrete Pressure Vessels

In his formal closing of the Conference on Prestressed
Concrete Pressure Vessels organized by the Institution of
Civil Engineers in London in March, 1967, Lord Hinton of
Bankside gave a brief history of the evolution of the de-
sign of pressure vessels for gas-cooled reactors[1].

Soon after World War II, designers in both the United
Kingdom, France, the United States and elsewhere realized
that the roles of the welded steel pressure vessel and the
concrete biological shield could be combined in a prestressed
concrete pressure vessel. However, cost studies in the
United Kingdom did not show the concrete vessels to be
cheaper, and the first gas-cooled reactor designed to produce
plutonium for the military program with electrical energy as
a byproduct used a steel pressure vessel, and was commissioned at Calder Hall in 1956.

At the same time, several similar designs aimed at a
civil nuclear program were nearing completion, and contracts
were placed by the Central Electricity Generating Board for
twin reactor stations at Berkeley (275 (MW(e)) and Bradwell
(300 MW(e)). As the Magnox reactor program evolved, gas
pressures and vessel sizes increased. Problems arose with
the site fabrication and inspection of the steel vessels,
and in 1959 the concept of an integral design of concrete
pressure vessel was being considered by Sir Robert McAlpine

93

and Sons Ltd. In this design, the reactor core and boilers
were contained within one vessel, the failure of which was
considered to be incredible (Figure 1). The safety problems
of the steel vessels, boilers, duct work and bellows units
were thereby eliminated and replaced by the relatively minor
problems associated with the inevitable penetrations of the
concrete vessel required for circulators, water, steam and
fuel access.

The first power station incorporating these features was
a twin reactor design of 660 (MW(e)) built by The Nuclear
Power Group Ltd, at Oldbury on Severn (Figure 1) and commissioned in 1967. A foil and mesh insulation developed in
conjunction with Darchem Ltd was used (Figure 5). The last
of the U. K. Magnox Stations, which was built at Wylfa (1200
MW(e)), used two internally spherical concrete vessels.
Insulation also included the element style developed in conjunction with Darchem Ltd (Figure 8).

The Magnox Stations were followed by Advanced Gas-cooled
Reactors (AGR), all of which use concrete pressure vessels.
Hinkley 'B' and Hunterston 'B' (1250 MW(e), seen in Figure 2),
were commissioned in 1976 and used a fibre insulation developed with Delaney Galley Ltd (Figure 7). Very similar designs using similar insulation are proposed for the AGR
stations approved by the government in 1978. The podded
boiler layouts of Hartlepool and Heysham, shown in Figure 3,
use element insulation.

It is interesting to note that presented with similar
problems, the French designers came to different conclusions,
and their first concrete pressure vessel contained only the
core. The G2 and G3 (40 MW(e)) reactors were built at Marcoule and commissioned in 1960. After reverting to steel
vessels for the Chinon Power Station EDF1 (70 MW(e)) and
EDF2 (200 MW(e)), a concrete vessel was again used to contain the core only in EDF3 (480 MW(e)), and this station was
commissioned in 1967. However, an integral design was chosen
for EDF4 (480 MW(e) commissioned in 1969) and subsequent gas-
cooled reactor stations(2).

In the French integral design, the vessel diameter is
kept to a minimum by positioning the boilers below the core.
The flow through the core is downward, thereby reducing the
surface area exposed to gas at the core outlet temperature.

Pumice concrete insulation (Figure 6) was installed in the
pressure vessels up to EDF4, but for Bugey 1, metallic insulation manufactured by Compagnie des Ataliers et Forges de
la Loire (CAFL) was used(3).

A downward flow core design, similar in many respects
to EDF4, was chosen by General Atomic for the Fort St. Vrain
Station (300 MW(e)). This helium-cooled, high-temperature
reactor was the first HTR to use a concrete pressure vessel
(Figure 4) and pressure and leak tests were completed in
1971(4). Multi-cavity vessels are proposed for subsequent
HTR's, and all designs have used or will use fibre insulation
for the low-temperature areas and a composite of ceramic
blocks and fibre for the high-temperature areas (Figures 9
and 10).

B.   Constraints on Liner and Concrete Temperatures

A mild steel membrane or liner on the inner surface of
concrete pressure vessels is required to ensure gas tightness.
This membrane is usually cooled on the concrete side by passing treated water or other liquids through a complex of pipes
welded to, or close to the liner. On the gas side of the
liner, the gas temperature can range between 200°C to over
700°C, and a thermal barrier, therefore, is required to restrict the heat loss to the cooling system.

In addition to the cooling pipes, anchorage features are
required to ensure that the liner and the concrete are effectively linked. After completion of concrete pouring, the pre-
stressing cables are tensioned and the liner is strained in
compression. Thereafter, and for all conditions of reactor
operation, the liner dimensions are effectively unchanged.
It is therefore necessary to ensure that liner temperatures
do not increase sufficiently to introduce unacceptable additional compressive stress or anchorage problems.

Temperatures and temperature gradients within the concrete also must be restricted. All these requirements can
be met in practice if the liner temperatures are limited to
60° to 80°C.

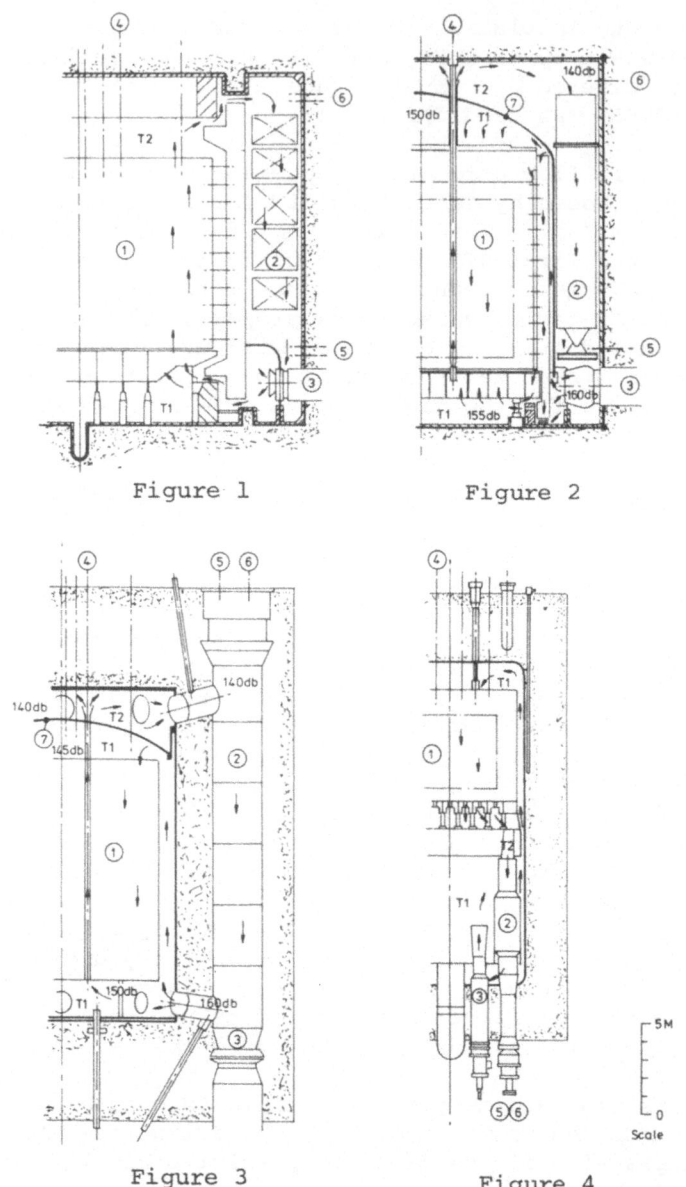

Figure 1   Figure 2

Figure 3   Figure 4

T1-Core Inlet Temperature. T2-Core Outlet Temperature
1 Core. 2 Boiler. 3 Circulator. 4 Standpipe.
5 Water Feed Penetration. 6 Steam Penetration. 7 Baffle.
Figures 1-4. Typical Reactor Layouts

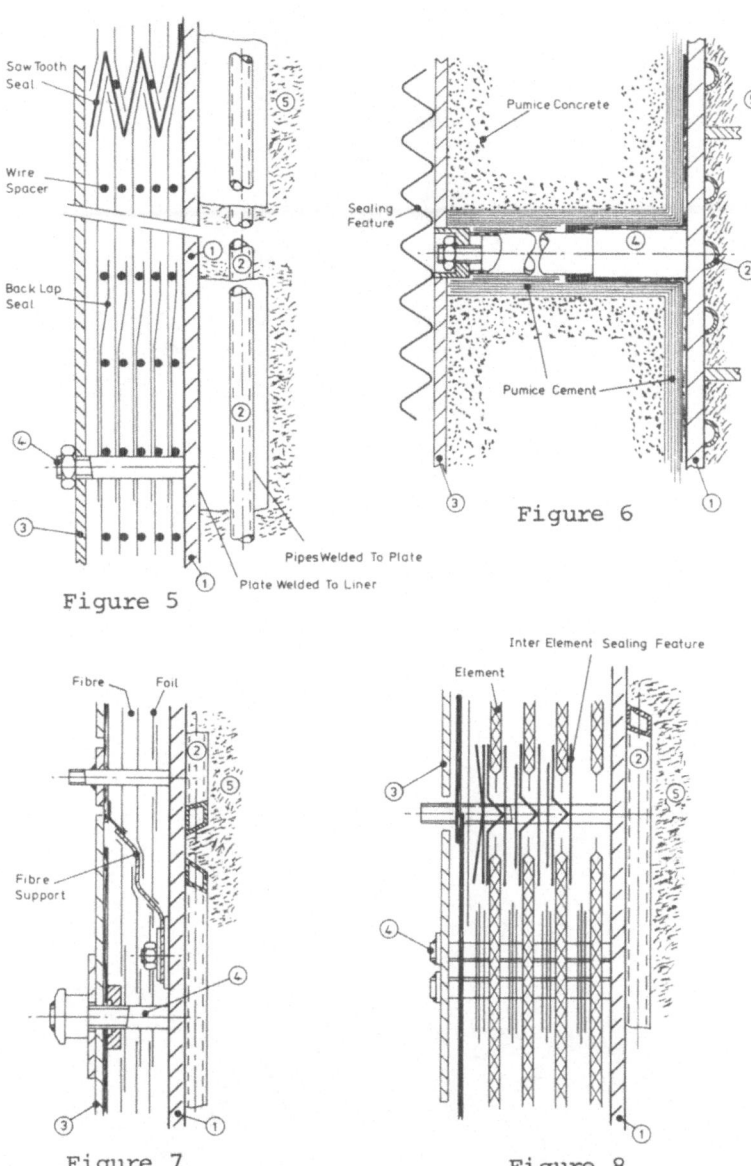

Figure 5

Figure 6

Figure 7

Figure 8

1 Liner. 2 Cooling Pipes. 3 Hot Face Cover Plates.
4 Stud or Strip. 5 Concrete. Number of insulation
layers reduced to show details.

Figures 5-8. Diagrammatic Sketches of Insulation Styles
and Cooling Systems

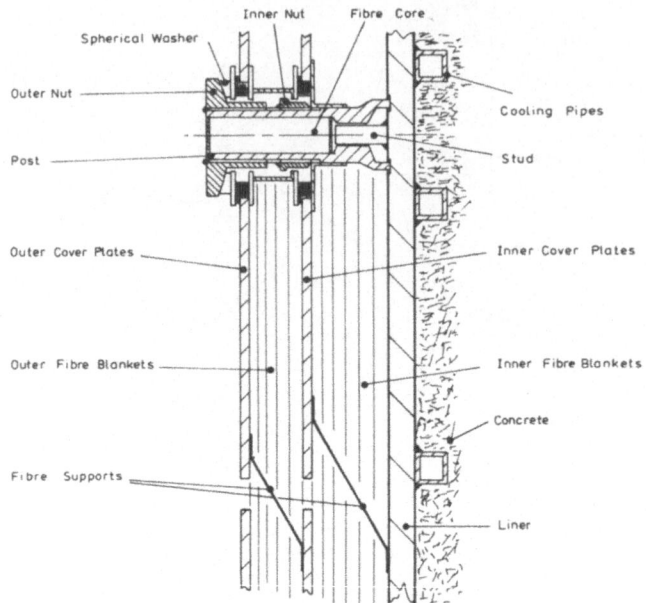

Figure 9

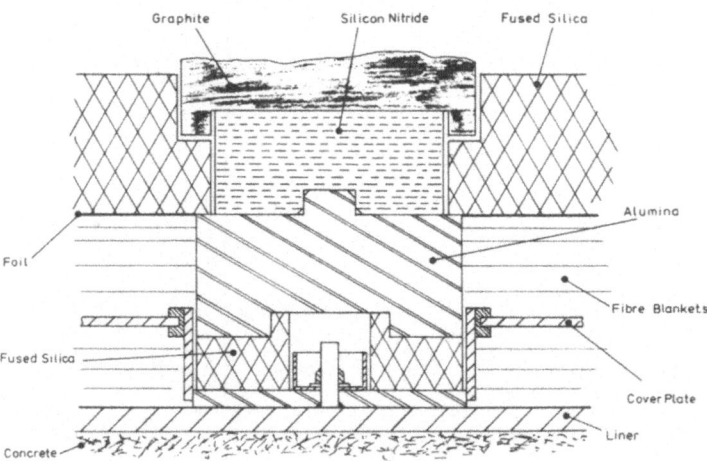

Figure 10

Figures 9 and 10.   Diagrammatic Sketches of Insulation
                     Styles - H. T. R.

C.   Cooling System

     The cooling system must be capable of removing the
heat passing through the insulation when the temperature of
the river, sea or atmosphere into which station heat is re-
jected is at its highest value. The design value of this
sink temperature will be fixed after studies of seasonal
and yearly variations. Two less obvious requirements are
to be met:

     1. Large local excess heat flow. If locally the heat
flow through the insulation is significantly above the design value, unacceptable temperatures will result unless the
maximum liner-to-cooling water temperature difference is low
(typically 5°-10°C). Liner temperatures would be expected
to peak midway between the cooling pipes, but it is not
normally possible to control the stud-cooling pipe align-
ments; therefore the worst configuration should satisfy this
requirement.

     2. Excess heat flow over a large area. In this situ-
ation, the liner temperature, which responds to the coolant
temperature, will be acceptable only if the design temperature rise of the cooling water is low (typically less than
20°C).

     Because of its influence on the reactor building pro-
gram, the cooling system design must be fixed very early and
in some areas, the details of the insulation and components
and hence the local thermal loading may not be known. Though
a vital factor in the liner temperature control, the cooling
system cost is usually a relatively small proportion of the
insulation cost.

     For these reasons, it is prudent to design the cooling
system very conservatively. Once installed, there is little
scope for modifications as it is largely inaccessible.

     Round, half-round(3) and square pipes have been used,
positioned and welded in a variety of ways (Figures 5, 6
and 7). Recent practice is toward a cooling system layout
as shown in Figure 7. In order to reduce the maximum liner-
to-cooling water temperature difference, the pitch of the
pipes and the thermal resistances between the pipes and the
liner are kept small. Continuous welds on both sides of the

cooling pipe, which itself is in direct contact with the
liner, is one solution. It is usual to have two independent cooling circuits and preferably, two 100% systems.
Critical liner temperatures should be considered with only
one circuit operating.

Penetrations present special problems. One solution
is to use twin circuits formed to spiral around the penetration with a pitch that is reduced toward the liner.
Alternately, the penetration can be surrounded by an annular
duct. If satisfactory distribution of the cooling water is
achieved, excellent heat removal characteristics result.
However, in this arrangement, the possibility of reactor
coolant gas leaking into the water is increased, as there
is only one barrier between gas and water, and there is no
secondary cooling system.

Floor, walls, roof and penetrations all require pur-
posely designed general pipe layouts sized so blockage is
not possible. The interconnection of these areas involves
a compromise between the thermal requirements and complex-
ity(5). During the commissioning phase, it is necessary to
demonstrate that the required flows are achieved in all circuits. Control of coolant chemistry is required to ensure
that corrosion problems are minimized(6).

D.   Insulation Specification

1. Types of thermal barriers. The thermal barrier
has been described as the Achilles heel of the concrete
pressure vessel. It is therefore not surprising that very
large development programs have been initiated in many countries to find the optimum solution(7-11). However, further
consideration must be restricted to designs that have progressed through the development stage to the building stage.
These fall into two main categories: active and passive.

Active designs isolate the liner from the main reactor gas environment by circulating low temperature gas
over the liner so the heat flow to it is small. Gas flow at
the required temperature under all conditions of reactor
operation (including shut-down) and over the whole liner
surface must be guaranteed. Heat transfer between the main
and cooling gas streams must be controlled. Although applied
to the Marcoule reactors G2 and G3, no recent design of

thermal barrier has used an active system, and it will
therefore not be considered further.

Passive designs use assemblies of material on the
liner to provide a high resistance to heat flow from the
main reactor gas environment. This insulation approach has
been used in all recent designs and therefore will be the
only one considered in greater detail in this chapter.

2. Cost optimization. The optimum design of the liner
insulation is obtained by a cost study involving the following factors (7):

a. cost of insulation and cooling system
   installed

b. cost of heat lost to the cooling system

c. cost of increased pressure vessel internal
   dimensions to accommodate insulation

d. cost of installation time

Diagrammatic sketches of typical designs for the
wall areas are given in Figures 5-9. Special areas such as
the floor, baffle and roof, which are discussed later, could
require different features (Figure 10).

It is relevant to note that the total cost of
liner insulation can be as high as 10% of the total station
cost.

3. General. In addition to satisfying the thermal
requirements, the insulation must be capable of surviving
30 years in a gas environment which subjects it to severe
temperature and pressure transients, and in which the acoustic noise levels are potentially damaging. In most areas,
radiation levels or restricted access make maintenance work
difficult, and in some areas, impossible.

E.   Working Environment

Typical gas conditions are summarized in Table I.

TABLE I

GAS CONDITIONS

(Figures give likely maxima or range)

Reactor Type	Gas	Inlet Temp. T_1 °C	Outlet Temp. T_2 °C	Pressure Bar $(10^5 N/m^2)$	Pressure Change bar	Pressure Gradient bar	Chemical Composition	Acoustic Level db
Magnox	Carbon Dioxide	160/250	360/410	10/25	0.2	5×10^{-3}	H_2O 500 vpm CO 1%	140
Advanced Gas Cooled	Carbon Dioxide	340	650	40	0.5	1.5×10^{-2}	H_2 250 vpm CO 2% CH_4 50 vpm H_2 100 vpm	120/165
High Temperature	Helium	250/350	700/800	40/60	20	1.0×10^{-1}	H_2O 2.5 vpm CO 25 vpm CH_4 20 vpm H_2 25 vpm	130/145

Special conditions apply as follows:

1. <u>Magnox reactors</u>. Compatibility requirements with
the fuel canning material are so severe that many conventional insulants can only be used if contained. In the
United Kingdom, only stainless steel insulants (Figures 5
and 8) have been used, although the EDF3 and EDF4 reactors
use pumice concrete contained by a membrane (Figure 6).

The Oldbury design (Figure 1) has been chosen to
illustrate the Magnox reactor; however, many liner features
of reactor EDF4 are similar to Figure 4.

2. <u>Advanced gas-cooled reactors</u>. Stainless steel canning material allows a wider range of insulants to be used,
and both ceramic fibres and stainless steel foils have been
chosen (Figures 7 and 8). When fibres are used, dust ingress into the circuit must be restricted to control activity levels, as these influence access problems.

Figures 2 and 7 give details of the Hinkley 'B' layout and insulation; Figures 3 and 8 show the podded boiler
design and insulation adopted for the Hartlepool Power
Station.

3. <u>High-temperature reactors</u>. The temperature level,
rate of pressure change, pressure gradient, etc. are now
particularly onerous. Furthermore, friction and wear characteristics for metal components are very sensitive to trace
impurities in the helium. However, because of the physical
properties of helium, natural convection and metal conduction
problems are less severe than in carbon dioxide.

The Fort Saint Vrain layout and the wall and core
support insulation are shown in Figures 4, 9 and 10.

F. <u>General Description of Insulation Styles</u>

In this section, only broad general principles affecting
thermal performance are described. Section III contains some
details regarding ways these and other considerations influence the design of insulation for the various ares of the
reactor.

1. <u>Insulant</u>. The insulant is the component of the insu-

lation assembly which provides the main thermal resistance.

a. Permeable insulants. One approach is to utilize the
fact that the gas coolant has a thermal conductivity several
orders of magnitude lower than conventional reactor materials (typically for CO_2 and helium 0.04 and 0.25 W/m K, and
40 and 20 W/m K for mild and stainless steel, respectively).
The problem is to reduce gas movement to an acceptable level
(a few meters per hour), and this requires materials of
very low permeability. There are two different classes,
both of which have achieved conductivity values of 0.1 W/m K
in high-pressure carbon dioxide:

> Isotropic materials - fibres and wires.
> Such materials simplify the design of
> corners and penetrations when the con-
> ductivity in directions normal to each
> other are important(12).

> Anisotropic materials - multi-foils.
> These foils are traditionally of
> stainless steel laid up so they pro-
> vide a very high resistance to gas
> movement normal to the liner. Parallel
> to the liner, even with internal seals,
> the resistance is several orders of
> magnitude lower(12,13).

b. Massive insulants. The second approach is to use
massive material which, in the working environment, has a
relatively low thermal conductivity (Pumice concrete k =
0.5 W/m K)(3). When used in carbon dioxide, because of the
high thermal conductivity relative to fibres and foils,
much greater thicknesses are required than for approach (a)
(Figure 6). However, when used in helium, the conductivity
differences are not so serious, and the good load-bearing
qualities and ability to operate at temperatures above 650°C
are unique advantages(4).

2. Retention details. In many designs, the insulant is
held on to the liner by a system of clamped cover plates
(Figures 5, 6, 7 and 8). The primary cover plate is massive
and ensures that the insulant is always in good contact with
the liner. The cover plates are supported on single or
multiple studs or strips that are welded to the liner. A

compromise must be struck between the stud stress and the
thermal shorting paths provided by the stud. Even after
very careful attention to the detailed design of the stud,
the ratio of stud to total heat conduction can range between
0.25 for wall insulation to 0.65 for floor insulation where
the acoustic environment demands a very robust design.

The primary cover plates are designed to allow expansion under all conditions and tolerances. To bridge the
gaps between primary cover plates, a secondary cover plate
is used (Figures 7 and 8). Further skins can be incorporated to increase hot face flow resistance, thereby improving
thermal performance and reducing the effect of surface pressure gradients(12,14). The whole assembly must, however, be
capable of withstanding depressurization rates as listed in
Table I. Additional layers and sealing features within the
insulant may be necessary to provide support or to increase
the resistance to gas movement (Figures 5, 7, 8 and 11).

II.   INSULATION THERMAL PERFORMANCE

In any insulation system, heat will be transported by
conduction, convection and radiation. Which of these processes is the most important varies from insulant to insulant.
In the massive insulants, it is the intention that the conduction process be of overwhelming significance, convection
and radiation being of little importance. For permeable insulants, however, the situation is more complex, with convection and radiation playing significant roles. Because of
the different nature of the heat transport processes involved,
it is appropriate to treat the permeable and the massive insulants separately.

Now the total heat passing through any insulation assembly
consists of two parts:

   that passing through the insulant;

   that passing through the studs of the
   insulation retention system.

The following sections concern themselves only with the
first of these, it being assumed that the stud contribution

is easily determined from a knowledge of the properties of
the stud, and the liner and insulation hot-face temperatures.

A. Permeable Insulants

As has been mentioned, the thermal performance of permeable insulants depends upon the gas trapped within the
insulant. In all systems developed so far, this trapped
gas is at the same overall pressure as that within the main
reactor vault. In this way, problems of insulation retention, consequent on pressure changes in main gas circuit,
are avoided. Equalization of the gas pressure between the
insulant and the vault is achieved by having a permeable hot-
face retention system (see Section I, F2).

As gas is driven around the reactor circuit by the circulators, pressure gradients over the face of the insulation
are created. At high pressure points, gas is forced through
the permeable hot face and into the insulation, eventually
returning to the main vault at points of low pressure. The
length of the flow paths for such gas movement may be appreciable. Superimposed upon this forced convection gas motion
is that produced by natural convective effects arising from
the non-uniform temperature distribution within the insulant.
It is thus apparent that within the insulation system the
pattern of gas movement can be quite complex, and dependent
upon the temperature distribution. At low overall gas densities, the transport of heat within the insulant is not
greatly changed by gas convective effects. At high circuit
gas pressures, however, when the coolant gas density is high,
appreciable convective effects can occur such that the temperature pattern and the consequent transfer of heat within
the insulation are substantially modified.

In permeable insulants, a further complication arises with
respect to heat transport by radiation. In order to have low
thermal conductivity, highly porous insulants are used. The
transfer of heat by radiation across the large pores in such
bodies occurs with ease, and for thin layers of insulating
material, may become the dominant thermal process.

1. General theoretical considerations. From the above
discussion it is apparent that the permeable insulants used
in reactors can be regarded as porous solids. Furthermore,
since the proportion of gas in these solids is large (in
order to cut down thermal conductivity effects) then the

porous solids are very open and have a porosity ε approaching unity.

The steady-state equations for fluid flow and heat transfer in such open porous solids are

the continuity equation:

$$\frac{\partial}{\partial x}(\rho_g v_x) + \frac{\partial}{\partial y}(\rho_g v_y) + \frac{\partial}{\partial z}(\rho_g v_z) = 0 \qquad (1)$$

the fluid momentum equations:

$$\frac{\partial p}{\partial x} = -\{\frac{\mu_g}{B_x} + C_x\, \rho_g |v_x|\}v_x \qquad (2a)$$

$$\frac{\partial p}{\partial y} = -\{\frac{\mu_g}{B_y} + C_y \rho_g |v_y|\}v_y \qquad (2b)$$

$$\frac{\partial p}{\partial z} + \rho_g g = -\{\frac{\mu_g}{B_z} + C_z \rho_g |v_z|\}v_z \qquad (2c)$$

the energy equation:

$$\frac{\partial}{\partial x}(k_x \frac{\partial T}{\partial x}) + \frac{\partial}{\partial y}(k_y \frac{\partial T}{\partial y}) + \frac{\partial}{\partial z}(k_z \frac{\partial T}{\partial z}) = \rho_g v_x \frac{\partial}{\partial x}(c_g T) +$$
$$+ \rho_g v_y \frac{\partial}{\partial y}(c_g T) + \rho_g v_z \frac{\partial}{\partial z}(c_g T) \qquad (3)$$

the fluid equation of state, usually the ideal gas law, i.e.,

$$p = RT\rho_g \qquad (4)$$

It is assumed, in the above equations, that the z axis is vertical, so the gravitation term $\rho_g g$ is present in the z direction momentum equation alone.

The parameters v_x, v_y and v_z represent the velocity of fluid movement around the insulant as a whole, and there-

fore may be termed the components of the macroscopic velocity
vector. '

    With regard to the form of the fluid momentum equa-
tions, these have been deduced from flow measurements carried
out on samples of both the fibrous and the stainless steel
foil-type insulants. Over the range of conditions experienced in reactor environments, the equations adequately describe the relationship between the pressure and the macroscopic velocity distribution. The fibrous and foil-type
insulants differ primarily in respect to the values of the
parameters B (permeability) and C (inertial flow resistance).

    Now because of manufacturing and erection tolerances,
any insulation system or insulation pack must be regarded as
a random structure. The B and C values therefore will be
random variables varying from build to build of nominally the
same structure. When determining B and C values, it is
therefore necessary to measure several samples of insulation
packs in order to find the average value and the upper and
lower bounds of the parameters.

    On examining the energy equation, it can be seen that
the proposed mathematical model of the porous solid allows
the transport of heat only by the mechanisms of conduction
and convection. The convective terms are those dealing with
macroscopic convection, dependent as they are upon the macroscopic velocity components v_x, v_y, v_z. Additional convective
effects, due to local (or microscopic) fluid motion, must be
included as an aspect of the thermal conductivity process
and allowed for in the parameters k_x, k_y and k_z. Likewise,
the local radiative transport of heat across gas spaces is
to be accounted for in the local conductive mechanism.

    2. <u>Steel foil-type insulants</u>. Examples of two kinds of
foil insulation systems are shown in Figures 5 and 8. It
can be seen that the insulants consist essentially of elon-
gated pockets of gas trapped within the steel assembly com-
prising the insulation structure. These gas pockets are not
effectively sealed from each other (see Section III, A3) and
gas movement through the body of the assembly at large can
occur. The system as a whole behaves as a porous solid, the
pores of which are the gas pockets.

    The parameters B_x, C_x, etc., given in Equations 2a, 2b
and 2c have been established by measurements for foil insu-

lation systems. Typical values of the parameters are given in Table II in which the x direction is taken as being perpendicular to the plane of the foils.

TABLE II

VALUES OF THE PARAMETERS B AND C FOR TWO FOIL SYSTEMS

Foil System	$B_x (m^2)$	$C_x (m^{-1})$	$B_y, B_z (m^2)$	$C_y, C_z (m^{-1})$
Foil and Mesh Type	1×10^{-12}	5×10^{10}	1×10^{-8}	0
Element Type	2×10^{-12}	1×10^{10}	1×10^{-4}	0

It can be seen that the pack resistance to fluid flow perpendicular to the foils is several orders of magnitude greater than that parallel to the foils. This implies that the porous solid to which the foil pack is equivalent is grossly anisotropic.

With regard to the thermal parameters k_x, k_y, k_z for foil insulants, a detailed treatment is given by Davidson (15) wherein expressions of the following form are deduced:

$$k_x = \alpha_x k_f + \delta_x k_g + k_g F_x (Ra) + \gamma_x T^3 \qquad (5a)$$

$$k_y = \alpha_y k_f + \delta_y k_g \qquad (5b)$$

$$k_z = \alpha_z k_f + \delta_z k_g \qquad (5c)$$

The various components contributing to the 'static' conductivity terms k_x, k_y, k_z are incorporated into the foregoing expressions in the following way:

a. Metallic and gaseous conduction are
    represented by the terms

$(\alpha_x k_f + \delta_x k_g)$, etc.

b. Local or microscopic natural convection
    within the gas pockets is given by a term
    of the form $k_g F_x$ (Ra), in which Ra is a
    Rayleigh number where

$$Ra = \frac{c_g \rho_g^2 \beta_g g d^4}{k_g \mu_g} \; (\frac{\partial T}{\partial x}) \tag{6}$$

c. Radiation across the gas pocket is
    represented by $\gamma_x T^3$.

Estimates of α_y, α_z, δ_y and δ_z can be made from
references dealing with porous solids(16 and 17). Because
of the dominance of k_x in the thermal conductivity tensor,
and because of the random nature of the insulation assembly,
it is usual to measure the various parameters involved in
the expression for k_x. This can be carried out in the following manner.

Measurements of k_x at vacuum conditions (when convection and gas conduction are negligible) over a range of
temperatures will enable x_x and γ_x to be determined. Thereafter, measurements of k_x at low gas pressures (when convection is negligible) will enable δ_x to be determined. In any
measurement, it is difficult to separate the effects of macroscopic and microscopic convection. The presence of one en-
tails the presence of the other. For this reason, the microscopic convection term $k_g F_x$ (Ra) is estimated from correlations applicable to natural convection in closed rectangular cells with isothermal walls. Such correlations are
given in references 18 and 19. It should be noted that
most foil systems are designed such that this microscopic
convective term is small and of little importance in the
overall insulation behavior.

3. Fibrous insulants. Examples of insulation systems
dependent primarily upon insulants of the fibrous type are
shown in Figures 7 and 9. In fibre packs, the influence of

microscopic fluid motion is negligible, all fluid movement
being effectively macroscopic.

Now from flow resistance measurements on fibre assemblies of use in reactor situations, it has been established
that the parameters C_x, C_y and C_z, given in Equations 2a,
2b and 2c, are zero. The fluid momentum equations are
therefore of the type

$$\frac{\partial p}{\partial x} = - \frac{\mu_g}{B_x} v_x \qquad (7)$$

which is the well-known D'Arcy equation.

It has also been established(21) that the parameter
$\bar{B}$, the permeability, is given by

$$B \; \alpha \; \frac{\rho_f^2 s^2}{\rho_i^2} \qquad (8)$$

Thus, B is inversely proportional to the insulation
pack density, ρ_i, for any given fibrous material. Most
commercially available fibres are available in bulk or blanket form. The bulk form is usually isotropic in its flow
resistance, but the blanket form shows a slight degree of
anisotropy. Typical values of permeability for some fibres
are given in Table III.

TABLE III

VALUES OF PERMEABILITY FOR BULK FIBRE MATERIAL
AT DENSITIES OF 100 kg/m^3

Fibre	B (m^2)
Rocksil	1×10^{-9}
Refrasil	1×10^{-10}
Triton	2×10^{-10}
Microquartz	1×10^{-11}

Regarding the thermal conductivity tensor, this is to all intents and purposes, isotropic. Thus, k_x, k_y, k_z are all equal, and given by an expression of the form

$$k = \alpha k_f + \delta k_g + \gamma T^3 \qquad (9)$$

The parameters α, δ and γ all vary with insulation pack density, for any given fibre. On comparison with Equation 5a, it can be seen that there is no microscopic convection term. As for the foil type insulation, the radiative transport of heat is accounted for by the term γT^3. As previously discussed, α, δ and γ may be deduced from measurements on actual sample packs at varying pressures and temperatures.

With regard to radiation, the term γT^3 applies only to thick insulation packs in which radiation from one of the boundary surfaces cannot reach the other by transmission through the pack. For thin insulation packs, for which this condition is not true, a more detailed analysis of the radiation component, based upon methods given for instance, by Viskanta(22), is required.

4. <u>Insulation seal characteristics</u>. Now as described in Sections I,F2 and III,B, it is usual practice to hold the insulation onto the liner by a system of cover plates attached to the hot face of the assembly. These cover plates provide additional resistance to gas flow into the insulant. From the point of view of the flow field, the hot face retention system may be looked upon as a leaky seal, the performance of which is given by an equation of the form

$$\Delta p = - \left\{ \frac{\mu_g}{B_s} + C_s \rho_g \left| v_s \right| \right\} v_s \qquad (10)$$

In this expression, Δp is the pressure drop across the seal, and v_s the mean fluid velocity through the seal. Since the performance of the seal depends upon small flow paths, it therefore follows that manufacturing and erection tolerances will affect the seal performance. In fact, as for the foil insulation system, the values of B_s and C_s in the above expression will be random variables. The average values of B_s and C_s and the upper and lower bounds must be determined from measurements on representative insulation packs with the hot face system in place.

The hot face seal is not the only type of flow re-
strictor used in insulation assemblies. The following seals
can be identified:

> Saw-tooth seal used in foil insulation
> (Figure 5)
>
> Inter-element seal used in foil insu-
> lation (Figures 8 and 11)
>
> Inter-blanket foil seal used in fibrous
> systems (Figure 7)
>
> Fibre support seal used in fibrous
> systems (Figures 7 and 9)

In addition, there are many other forms of sealing
insulation packs in special areas (e.g., to penetrations
through the packs). All the seals have flow-pressure drop
relationships of the form of Equation 7.

5. Boundary conditions. In order to solve the set of
Equations 1 to 4 for any insulation assembly, not only must
the material properties described above be determined, but
also the thermal and fluid flow conditions on the boundary
of the system must be given.

With respect to the flow field, the necessary boundary condition is the pressure distribution (including the
static head component) over the surface of the cover plates.
In most cases, this pressure field is determined from flow
measurements on scale models of the reactor circuit (see
Section V,B).

With regard to the thermal boundary conditions, the
situation is more complex. In those areas in which gas flows
into the insulation, the cover plate system will be at the
temperature of the mainstream gas. Wherever gas emerges from
the insulation, however, the emergent gas is cooler than the
mainstream gas and the coverplate temperature is indeter-
minate. The usual practice is to assume that the coverplate
system is everywhere at the mainstream gas temperature.
This ensures that the heat passing through the insulation is
overestimated.

Having determined the set of Equations 1 to 5 for
the insulant, and Equation 10 for the seals, and having produced the requisite data and boundary conditions for use
with these equations, then theoretical estimates can be made,
usually by digital computer, of the velocity and temperature
distributions within reactor insulation installations. Examples of such calculations and their comparison with experimental work are given in references 15,21,23.

6. <u>General discussion of insulation thermal performance</u>.
In order to illustrate different facets of the behavior of
the two main types of permeable insulants, consider the pack
of insulation ABCD shown in Figure 15a, in which the hot
face is AD, the cold face BC. Assume that the hot face is
permeable and covered by a seal of the type discussed in
Section I,F^2. The cold face, being attached to the liner,
is, of course, impermeable. The ends of the pack are assumed
to be perfectly sealed and perfectly insulated. Along the
face AD, it is assumed that there is an imposed pressure distribution (in addition to any static head) this pressure distribution being linear, with the highest pressure at the
point A. The face AD is at a uniform temperature while the
liner face BC loses heat to a constant temperature cooling
system through a medium having a uniform heat transfer coefficient.

The pack ABCD can represent the roof, wall or floor
of any reactor installation, depending upon the orientation
of the pack with respect to the vertical. In floor and roof
configuration, natural convection will be insignificant (the
static head along the face AD being constant), but for wall
assemblies, natural convective effects cannot be ignored.

If there is no circulation of gas within the pack
ABCD, heat would pass through the system by conduction, and
the heat, Q_{st}, removed from the assembly over the cold face
would be uniform. Gas circulates within the insulant, however, as a result of forced, or natural convective effects.
This gas movement modifies the heat flow distribution within
the pack, and in particular, the heat, Q_{dy}, extracted from
the cold face. The distribution of Q_{dy} will not be uniform
over the face, BC.

A Nusselt number, Nu, can be defined at any point on
the face BC as:

$$Nu = \frac{Q_{dy}}{Q_{st}} \tag{11}$$

In situations in which no gas movement occurs, Nu is unity over the face BC. In dynamic cases (i.e., when gas moves within the pack) this is no longer true. The extent of the departure of Nu from unity represents the local effect of gas convection on heat transport within the system.

The modification, by gas convection, of the heat flow distribution within an insulation assembly has two important consequences from the point of view of reactor design:

a. There is an increase in the overall heat transport through the insulant. This is represented by a Nusselt number, $\overline{Nu}$, averaged over the face BC, the value of which is greater than unity. Any cooling system designed to remove only the conducted heat must be augmented if significant values of $(\overline{Nu}-1)$ arise.

b. Any non-uniformity of the distribution of Nu over the face, BC, implies a non-uniform liner temperature. It is necessary to ensure that the maximum liner temperature does not exceed limits imposed on the system from a consideration of liner and concrete properties. It should be noted that laboratory testing of insulation specimens is usually restricted to rigs of overall dimensions smaller than those in reactor installations. Considerable extrapolation is needed if measured Nusselt number distributions are to be translated into results directly applicable to reactor geometries.

In references 21 and 24, insulation systems based on foil and fibrous assemblies, respectively, are discussed. For natural convection phenomena, it is shown that:

$$(\overline{Nu} - 1) \ \alpha \ (\frac{RD}{H})^{1/2}, \qquad \text{for large RD/H} \tag{12}$$

and

$$(\overline{Nu} - 1) \ \alpha \ \frac{RD}{H} \ , \qquad \text{for small RD/H} \tag{13}$$

where

$$R = \frac{C_g \rho_g^2 \beta_g \, gB_x \, D\Delta T}{\mu_g k_g} \tag{14}$$

is the overall pack Rayleigh number.

In the foregoing expressions, the following points should be noted:

a. for fibrous material the parameter B_y can be used in the definition of R since B_x and B_y are not very different.

b. for foil material, the relationship will be only approximate since no mention is made of the inertial resistance term

$$C_x \rho_g |v_x| v_x$$

shown in Equation 2a.

With respect to forced convection, in fibrous material(21), it is shown that:

$$(\overline{Nu} - 1) \; \alpha \; \sqrt{F} \; , \qquad \text{for large F} \tag{15}$$

and

$$(\overline{Nu} - 1) \; \alpha \; F, \qquad \text{for small F} \tag{16}$$

where

$$F = \frac{C_g \, \rho_g \, B_x \, D^2}{\mu_g \, k_x \, H} \left(\frac{\partial p}{\partial z}\right) \tag{17}$$

The term $\partial p/\partial z$ in the expression for F is the magnitude of the pressure gradient along the hot face of the pack. Similar relationships to the above can be expected to hold for forced convection in foil systems.

In order to minimize problems arising from excessive

gas convection, it is usual to run reactor installations at
small values of the parameters F and RD/H, with modest increases in Nu. For such values of the parameters, illustrations of the variation of Nu along the liner for packs of
fibrous and foil insulants are shown in Figures 15b and 15c,
respectively. The illustrations are for packs in which the
relative dimensions are those of reactor wall configurations
and are typical of both forced and natural convective situations.

      As can be seen from Figure 15b, for fibrous insulants
there is an entrance and an exit region, in both of which
gas flow in directions perpendicular to the pack hot face
occurs, with a consequent pronounced effect on local Nusselt
number. There also is a central uniform region in which gas
flows only in directions parallel to the hot face. In this
central region, the Nusselt number is unity, the heat apparently passing through the pack by the mechanism of conduction
alone.

      For foil insulants, the situation is quite different.
The entrance and exit regions merge; no central uniform
region exists, and gas flow in transverse directions is
present everywhere in the pack. In fact, it may be demonstrated by dimensional analysis that tall foil packs composed of material having grossly anisotropic flow resistance
tensors behave in the same way as very short packs of isotropic material.

      Now it may happen that the maximum Nusselt number on
the reactor floor, wall or roof leads to too high a liner
temperature. Modifications to the insulation assembly are
then needed to bring the liner temperature within acceptable
limits. This can be achieved by the use of surface seals,
or longitudinal and transverse seals of the kind delineated
in section II,A4. Furthermore, it is advisable to discon-
nect the roof, walls and floor from each other, both for
ease of erection and to control forced convection that would
otherwise occur from the high to the low pressure regions of
the reactor. Again, transverse seals are employed to bring
this decoupling about. In various parts of the reactor, components penetrate the insulation and seals must be provided
to prevent the influx of gas into the assembly through such
local irregularities.

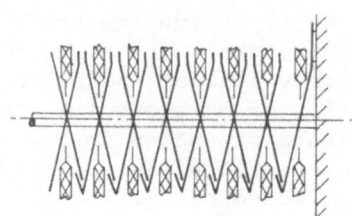

Figure 11.   Inter-element Seal

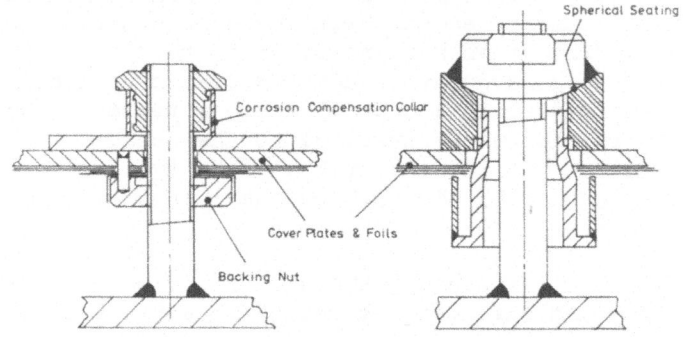

Figure 12.   Stud Detail (1)   Figure 13.   Stud Detail (2)

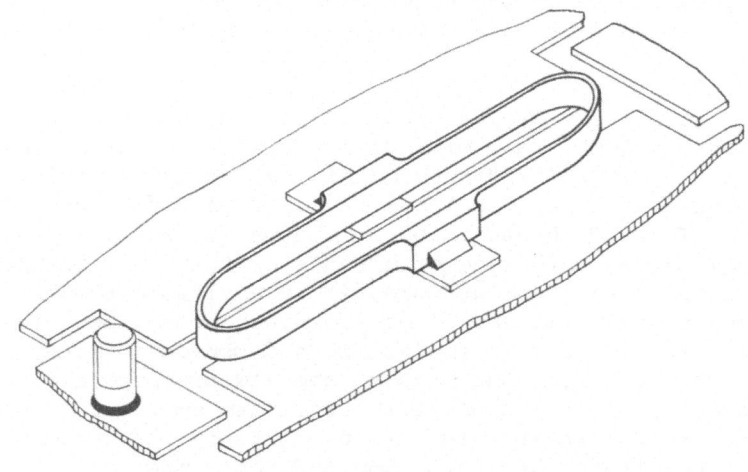

Figure 14.   Secondary Retention Loop

INSULATION DETAILS

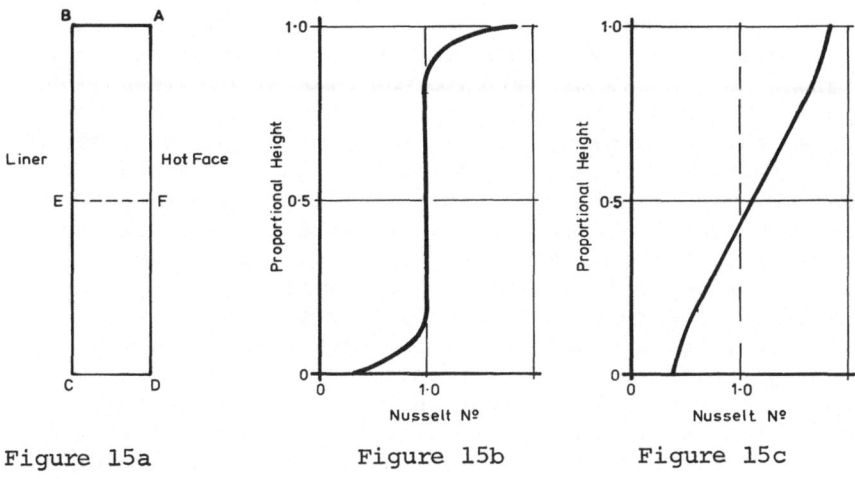

Figure 15a   Figure 15b   Figure 15c

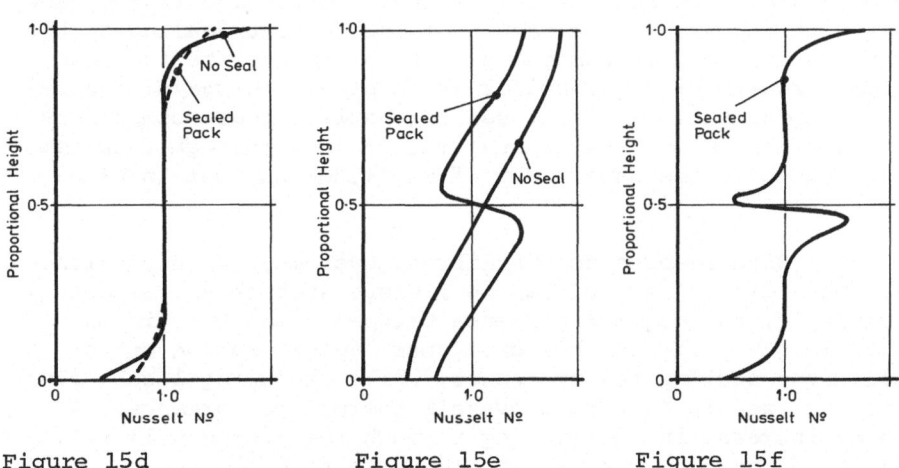

Figure 15d   Figure 15e   Figure 15f

VARIATIONS IN LOCAL NUSSELT NUMBER ARISING FROM HEIGHT AND
SEAL EFFECTS

To illustrate the effects of seals, consider the pack ABCD of Figure 15a, and assume that natural convection is not present (that is, the pack represents floor or roof insulation). Consider the situation in which the surface seal resistance is low. Under the action of pressure gradient along AD, gas is driven into the pack at A, flows along the path ABCD, and eventually leaves the system at D. Now the flow resistance along the path ABCD is composed of three parts, namely, the insulation resistance in the lengths AB, BC, and CD. The amount of gas driven along the path ABCD will depend upon the pressure drop between A and D and the total resistance along the flow path. For foil systems, the major resistance to flow lies in the direction perpendicular to the foils, the resistance parallel to the foils being small. Thus, the lengths AB, CD provide most of the resistance to gas flow around the circuit. It follows that any disruption of the pack assembly that reduces the through thickness resistance at any two or more separate points will lead to a significant increase in the amount of gas entering the insulation assembly, with a consequent deterioration in thermal performance.

Reduction of the through thickness resistance may be due to damage to the insulation pack. More importantly, however, it can arise as a result of inefficient sealing of the pack to any reactor component that must penetrate the insulation assembly; e.g., boiler steam pipes, charge/discharge standpipes, etc. In fact, sealing between these components and the insulation system must result in a through thickness resistance of the same order of magnitude as that in plane pack areas.

With respect to the fibrous systems, the situation is quite different. Since the fibrous systems are almost isotropic, the major resistance between A and D is in the path length, BC. In this case, pack deterioration or inefficient sealing to components, while producing local effects will not result in a poor overall thermal performance. A major increase in gas flowing through the system will only arise if the flow resistance along the path length BC is reduced. Thus, the occurrence of longitudinal gaps, parallel to the cold face over most of the length, BC, is necessary if a major insulation breakdown is to appear. Gaps of this nature may occur as a result of loss of fibre resilience (Section IV,B2) in roof assemblies unless special precautions are taken in the assembly design.

When the surface seal resistance is appreciable, then the total resistance along the path length, ABCD, is composed not only of the pack resistance but also the additional resistances at A and D due to the seal. In order to affect the overall performance of the pack, the seal resistance must be of the same magnitude as the insulation resistance from A to D.

In the case of the fibrous insulants, seals that are not resistive enough to affect overall performance do, however, produce local effects that are of importance. Thus, as illustrated in Figure 15d, the sealed pack will still have a central region in which the gas flow is the same as in the unsealed case. Over the entrance and exit regions, however, the gas velocities into and out of the pack are reduced, with a consequent reduction in the effects of local gas convection. In particular, the maximum Nusselt number in the system is reduced. At the same time, the extent of the entrance and exit regions is augmented to provide the same total flow of gas through the system. Effects similar to those described above are produced by interblanket foil seals as used in the insulation system shown in Figure 7.

In practical reactor installations as mentioned above, transverse seals; that is, seals passing through the thickness of the insulation assembly from the hot face to the liner, frequently are used. Examples of such seals are the sawtooth and the inter-element seals of Figures 5 and 11, and the fibre support components of Figures 7 and 9. The effect of these seals on gas flow within the insulation structure obviously depends upon the seal resistance. Seals of very low resistance will leave the flow pattern unchanged, while those of high resistance are likely to produce significant effects. To illustrate how the thermal performance of insulation assemblies is affected, consider a perfect through thickness seal introduced into the pack ABCD of Figure 15a, along the central plane, EF.

For foil insulants, the total resistance along the path, ABEF, will be the same as for the path, ABCD, in the unsealed case (the resistance between E and F replaces that between C and D). The pressure difference between A and F will, however, be only half that between A and D. The gas flow into the system, therefore, will be substantially reduced, in both the subpacks ABEF and EFCD. Nusselt number

curves of the form given in Figure 15c will result in both
these subpacks, the departure of the extreme Nusselt numbers from unity being approximately halved. At the seal,
EF, conduction of heat from the lower to the upper pack,
across the seal, will occur. The resulting Nusselt number
curve for the pack as a whole will be as shown in Figure 15e.
It can be seen that the introduction of the through thickness seal has substantially reduced the maximum Nusselt number in the system.

     For fibrous insulants, the situation is quite diff-
erent. Thus, the flow resistance along the path ABEF will
be half that along the path ABCD (the length BE is one-half
the length BC). The pressure difference, A to F, is half
that between A and D. It is therefore apparent that the gas
flow into the subpacks ABEF, EFCD is the same as that into
the unsealed pack ABCD. The subpacks, therefore, will behave in the same way as the original unsealed pack, and will
have Nusselt numbers of the same form as those given in Figure 15b, the extreme Nusselt numbers being the same for both
configurations. The resulting compound curve for the pack
as a whole will be as shown in Figure 15f. As can be seen,
through thickness seals for fibrous insulants will, in general, not affect maximum Nusselt numbers to any great extent.
As previously indicated, however, the value of $\overline{Nu}$ for the
distribution shown in Figure 15b is greater than unity. It
is to be expected therefore that the value of $\overline{Nu}$ for Figure
15f will be greater than that for Figure 15b. This implies
that the total heat rejection to the cooling system for the
fibrous insulation pack with transverse seal is greater than
that for the unsealed case, while the maximum Nusselt numbers
in the two systems are approximately the same. Thus, the
introduction of the fibre support components of Figures 7 and
9 into insulation packs have a deleterious effect upon the
thermal performance of the assembly.

B. Massive Insulants

     Wherever massive insulants are used as the primary insu-
lant in reactor installations, it is the intention that the
conduction process shall be by far the dominant means by which
heat is transported throughout the system. If this is achieved
computer programs for the solution of the conduction equation
will enable temperature distribution within the reactor insulation assembly to be assessed. For such an assessment, the

insulant data required are the thermal conductivities.
These are relatively easy to provide from laboratory measurements on small samples carried out in the coolant gas at
the requisite reactor pressures and temperatures.

If conduction is to dominate the thermal behavior, it is
necessary to ensure that small gaps are not present within
the assembly, such that gas movement can take place from
the main vault into such gaps between the insulant and the
liner. Convection phenomena occurring in these small gaps
can completely destroy the integrity of the insulant. The
major problem in installations using massive insulants is
one of ensuring the mechanical stability of the structure
such that small flow passages do not occur.

In some installations, massive insulants are used in a
subsidiary role; for example, in the core support insulation
of the Fort Vrain reactor (Figure 10). In such cases,
gas movement around the massive insulant can arise. Assessment of the thermal characteristics of such assemblies must
include allowance for the convective effects of these gas
flows.

C. Penetrations

For a variety of reasons the concrete vessel of gas-
cooled reactor systems is pierced by a number of penetrations. Typical of these are the control and charge/discharge
standpipes, instrument penetrations, and penetrations carry-
ing boiler steam pipes (see Section III,D4).

The penetrations are in many instances, open to the main
reactor vault, and therefore, like the major portion of the
reactor containment, require insulation and cooling systems.
Estimates of the performance of insulation assemblies in such
penetrations are difficult because of the boundary conditions
to which the insulant is subject.

For vertical configurations, the situation arises in which
the penetration is filled with gas that is hot at the bottom
and cold at the top. This results in the generation of severely turbulent, buoyancy-dominated, natural convection that
becomes the controlling factor in the transport of heat along
the penetration. In such circumstances, it is usual to resort to full-scale testing of the penetration and its components, for instance in the case of standpipes (Section IV,A3).

For horizontal penetrations, natural convection phenom-
enoma make themselves felt in a different fashion. Complex
patterns of gas flow can arise (27), which subject the full
length of horizontal penetrations to elevated gas temperatures. Effects of this kind cannot be ignored in the design of insulation and cooling systems.

III. GENERAL DESIGN CONSIDERATIONS

It is important to realize during the design and development phase that:

the surface to be insulated (the liner) is not
ideal but is a welded structure with surface
irregularities;

the special areas constitute a significant
proportion of the total area of the liner;

acoustic problems can be formidable.

Many of the designs proposed overlook these considerations and therefore have no real reactor relevance.

A. Insulant Problems

The insulant is the key factor in the thermal performance of any design. It must achieve and maintain the design performance throughout the service life of the reactor.

1. Fibre damage criteria. The use of fibres in a hostile
reactor environment is novel, and damage criteria must be
established(12). Because of the high surface-to-volume ratio,
stringent tests are required to ensure that no chemical problems arise and that the fibre is compatible with its surroun-
dings. It is also necessary to establish that properties are
not influenced by water or oil ingress, or by radiation.

In the reactor, the containing components will re-
spond to the acoustic environment so the fibre is in contact
with a moving surface. This movement must not cause dusting
or deterioration of the fibre.

2. Fibre resilience and support. To retain the fibre,

it is clamped between the cover plates and the liner. At
temperature, this blanket-clamping pressure reduces in a
manner that must be understood so conditions at the end of
30 years can be predicted. Fibre support could be lost if
blanket pressure falls to zero, although methods of bonding
the fibre to the liner are under consideration, and rig work
to date would suggest that this problem should not be difficult to resolve. Triton Kaowool and Quartz et Silice fibres
have been used in Hinkley and Fort Saint Vrain.

    3. Foil requirements. Early arrangements of foil in-
volved a multi-layer assembly spaced by a wire mesh and
clamped to the liner by a cover plate stud arrangement(12-14)
(Figure 5). Because of foil overlaps, working clearances,
irregular liner surface, freedom to slide, depressurization
rates and other considerations, it is not possible to achieve
isolated cells within the assembly. For example, at Oldbury,
the nominal thickness of the insulant was 50 mm, and this
contained 30 sheets of foil, 0.10 mm thick, spaced by wires
1.22 mm in diameter. Thus, a void of 12 mm was left, and
this was large enough to give a very low resistance to gas
movement in the plane of the foils. To increase this resistance, a back lap seal was formed by a suitable laying-up procedure, and sawtooth seals were spaced as appropriate through
the pack (Figure 5) to reduce the maximum Nusselt numbers, as
described in Section II, A6.

    A compromise must be struck between the foils' being
thick enough to slide over its neighbors without buckling,
and being thin enough to reduce the thermal conduction area
to an acceptable value. Stainless steel is used for both
thermal and corrosion reasons.

    A development of the Oldbury style of foil insulation
was used at Wylfa, and subsequent reactors were designed by
the British Nuclear Design Company(14). Elements are formed
by welding two foils together along the four edges after insertion of a dimpled foil-spacing feature. The element must
be pierced to allow pressure change to be accommodated (Figure 8).

    Inter-element sealing is required to control gas
movements external to the elements. This movement is possible
because the element faces are only in limited contact with
their neighbors, and gaps are left between edges to allow for
differential expansion.

The element arrangement has particular advantages when the penetrated areas are to be insulated. Furthermore, because fewer individual components must be handled, erection problems are eased. Compared with the Oldbury style, more foil material is used with a consequent thermal disadvantage of greater heat flow in the plane of the foils. A typical effective conductivity in this plane would be 20 times that in the normal direction.

The insulation developed by CAFL uses a mattress of gauze woven from stainless steel wire approximately 0.2 mm in diameter, contained between foils, as its basic module. Multi-stud attachments are used to support the number of modules required to provide the design thermal resistance.

4. _Massive materials_. Pumice concrete has been used by the French in their Magnox reactors EDF3 and EDF4(_3,25_). However, as this was not compatible with the fuel element canning material, it was contained by membranes to ensure that dust particles were not circulated by the main gas stream. An impermeable membrane was used in EDF3 and as initially designed, the pumice concrete operated at atmospheric pressure(_3_). However, problems with integrity under transient pressure conditions led to the use of a permeable membrane on EDF4, and the insulant was then at full reactor gas pressure.

The thermal conductivity in the reactor environment must be established. It is then necessary to mount the block on the liner in such a way that gas flow between the block and the liner, and between block and block, is so restricted that the required thermal performance is achieved (see Section IIB). The sealing feature incorporated must be capable of accommodating differential movements (Figure 6).

Ceramic blocks are used in the Fort Saint Vrain High Temperature Reactor for part of the insulation of the core support floor that is located downstream of the core. Ducts convey gas, with the possibility of streaks at temperatures as high as 1085°C, to the steam generator modules. In this area, fibre insulation is contained by cover plates on top of which the ceramic blocks are mounted (Figure 10). Details of the work completed to establish that the material will survive in the thermal, acoustic and pressure environment of the reactor are described by Jones and Hedgecock(_26_).

B. Hot Face Details

Typical hot face details are shown in Figures 5 - 10,
and further information is available (3,12,13,14,26,27).

1. Primary cover plates. These plates provide the main
insulant support feature. The insulant is compressed to a
predetermined pressure to ensure good support at the end of
reactor life; creep of cover plates can present problems.
The primary and secondary cover plates protect the insulant
from mechanical damage during construction and acoustic damage during operation.

2. Secondary cover plates. Gaps between the edges of
primary cover plates are required to allow for differential
expansion. These gaps are covered by secondary cover plates
usually similar in size to the primary plates, but mounted in
a staggered array.

In addition, foils frequently are incorporated to
provide greater resistance to hot face gas flow and to ensure
that particles of insulant do not escape into the main gas
stream. Problems of sliding usually can be eased by surface
treatment of components, or by spacing washers or rubbing
strips of suitable material.

C. Retention Details

The insulant always must remain in good contact with the
liner. If gaps are allowed to develop, gas flows can take
place that effectively neutralize the thermal resistance of
the insulant (Section II,A6).

1. Primary retention. In many designs, cover plate details are retained by studs welded to the liner. A compromise is required between massive studs operating at low stress
levels and slender studs restricting the heat flow conduction
path. For carbon dioxide-cooled reactors, the allowable
cross-section area of steel in the studs is limited to typically 1/300 of the module size. Bi-metallic studs achieve
the simplicity of site-welding mild steel on mild steel and
the advantage of the lower thermal conductivity of stainless
steel. Hollow studs are frequently used (Figures 6 and 9).

The nut detail can be of the simple form used in
Oldbury (Figure 5), which is locked by welding (12). More

complex forms clamp the cover plate system against a backing nut (Figures 7 and 12). Provision then must be made for interface corrosion to ensure that unacceptable local strains cannot arise. A further development would be a spherical seating arrangement (Figure 13) which would allow the angular setting of the cover plates to be adjusted for liner imperfections and stud misalignment (13). For a multi-stud retention system, thermal expansion requires that deflection or sliding of components can take place. If deflecting studs are used, they will be grouped near each other and attached to the cover plate in a manner that provides maximum stud length. The sliding alternative requires first, an extensive test program to substantiate the materials used; and second, a well-defined assembly procedure to ensure stud and component alignment, and that loads are to specification, and that these are maintained in reactor operation.

2. <u>Secondary retention details</u>. Single stud attachment requires that provision must be made for a secondary retention system to provide cover plate location if the stud fails. In Figure 14, a loop system is shown. This loop system allows differential movement of the cover plates, but is very stiff in the direction normal to the plates.

D. <u>Special Areas</u>

The insulation described in the subsections C2 and C3 would be used on the plain wall areas. Other areas, which can form a significant proportion of the total area to be insulated, are described in this subsection.

1. <u>Floor</u>. In Magnox and AGR systems (Figures 1, 2, 3), low gas temperatures ease the design problems in this area. However, two special features arise. The construction program can dictate that floor insulation is installed early, and further work is carried on thereafter. The design must therefore have a load-bearing ability not required for other areas. Furthermore, being close to the circulators, the noise level is high, and for these reasons, a robust design is required. An additional problem demanding particular attention to sealing details arises wherever the surface pressure gradients are large. Insulant support is not a serious problem, but the influence of water (from boiler leaks) or oil (from circulator leaks) must be assessed. Radiation levels are high in the undercore zone.

Following the lead of Fort Saint Vrain, future HTR designs(28) propose a downward flow core so the charge machine can operate at the low gas temperature, T_1. However, the reactor core/boiler arrangement is now closer to Figure 3 than to Figure 4, and the insulation problems are considerable, due to core load-bearing requirements, temperature level, pressure gradients and radiation levels. Circulators are remote from the floor, so acoustical problems are eased.

2. Roof. The roof is highly penetrated with fuel and control rod standpipes. With an upward flow core during fuel charge/discharge, the roof above an empty channel could be exposed to a high velocity, low-temperature gas stream that results in severe temperature and pressure gradients. Furthermore, insulant support is critical because any movement away from the liner would enable high-temperature gas to flow between the insulant and the liner with quite unacceptable consequences.

3. Baffle. The high and low temperature gas streams in an AGR (Figures 2 and 3) are separated by a large internal steel pressure vessel known as the baffle(27). Because of problems of differential movement and cost, the baffle is of mild steel, and requires insulation to maintain it at a temperature of around 340°C. Strictly, the baffle is not part of the liner. It is, however, one of the critical pressure components of an AGR, and is included here for completeness.

Unique problems of insulation arise because the center of the baffle is highly penetrated and the normal water-cooling system cannot be used. The lower surfaces of the baffle are cooled by low temperature gas (T_1) which subsequently flows into the core. Bleed flows cool the nozzles.

4. Penetrations. Standpipe penetrations are effectively high-aspect ratio pipes penetrating the roof slab. The diameter is too small to allow man access; therefore, the insulation is assembled onto an inner tube prior to insertion into the liner. It is essential to ensure that no uncontrolled gas movements are possible between the insulation assembly and the liner, and absolute seals are required. The bore of the standpipe, defined by the fuel element/grab size, for the Magnox and AGR systems, was 230 mm and 300 mm, respectively. However, in recent designs of HTR, it was proposed to discharge the fuel element and its surrounding graphite

brick as an assembly. Bores of 700 mm will be required.

Penetrations through the walls of the reactor can
range in size from 2 m for circulator housings, down to
400-200 mm for water steam and instrumentation pipes. Steam
penetrations can enclose pipework that is subject to large
movements due to boiler displacements at operating temperatures. Some form of gas seal at the mouth of the penetration
becomes necessary, which must allow for both radial and vertical movements to take place(27).

5. Gas ducts. A development of the single-cavity pressure vessel involves the use of boiler pods connected by
large-diameter gas ducts to the core cavity (Figure 3). These
ducts are large enough to allow man access, but the insulation
requires special features because of the very high surface
pressure gradients that result from the gas accelerations up-
stream and downstream of the duct.

The various solutions to this design problem involve:

a. an inner duct that is sealed at one end
    and open at the other, isolating the
    insulation from the pressure gradient
    but allowing the insulation to respond
    to circuit pressure changes.

b. the use of an impermeable membrane on the
    insulation hot face that is so formed that
    axial and circumferential movements can be
    accommodated.

c. a special form of multi-foil or element
    insulation in which the layers nearest
    the hot face have flow resistances that
    reduce the pressure gradients on the
    effective insulant(9).

d. the use of radial sealing features to
    minimize the flow in the axial direction(9).

IV. DESIGN SUBSTANTIATION

A. Thermal Performance

When concrete pressure vessels were first proposed, the
use of insulation in gases at high pressure, in the sizes
required, and under the temperature and pressure gradient
conditions of gas-cooled reactors, was quite novel. However,
quite early in the development program planned to provide
designers with thermal performance data, it became clear that:

large, high-pressure test rigs,

and a sophisticated mathematical model

were essential to progress. The cost and complexity of test
rigs to investigate all reactor insulation details would be
prohibitive, and it was therefore necessary to understand the
basic heat transfer processes and establish thermal performance prediction methods. With this model, it has been possible to predict the temperature distribution and heat flow
for a wide range of boundary conditions under both natural
convection and forced-flow situations.

1. Natural convection. Initially in the United Kingdom,
testing was carried out in small rigs similar in arrangement
to the standard conductivity rigs used for building materials,
but adapted so that testing at high gas pressure was possible.

From the analysis of these tests, it soon became
clear that larger, higher-pressure rigs were required, and
foil and wire specimens of 2 m and 5 m height eventually were
tested.

Other investigators were confronted with similar
problems. For example, the French Grand Megabidon test vessel was 18 m high and capable of operating at gas pressures
up to 60 bar. A test loop was built in Germany in which
large specimens could be tested under natural convection and
pressure-gradient conditions(29).

The initial United Kingdom problem was to refine the
details of the foil and wire insulant so the thermal performance required for the Oldbury design could be achieved.
Numbers of foils, spacing of foils, clamping loads and sealing features were some of the variables investigated. Most

of the tests were carried out in high-pressure carbon dioxide
with only a limited helium program, planned to substantiate
the significance of Rayleigh number correlations suggested
by the theoretical work of Section II.   Insulant permeability
and hot face flow resistance also were required for interpretation of the data.   The procedure developed involved the
assembly of multi-module rigs designed so the permeability of
the rig assembly with flow normal and parallel to the liner
could be measured.

Subsequent to this and with the AGR application in
mind, a much broader investigation of granular and fibrous
materials was completed and a general correlation derived.
Test work had established that permeability values were independent of pressure and gas, which allowed the majority of
the specimen tests to be carried out in atmospheric pressure
air, thereby greatly simplifying the test work.

2.   Surface pressure gradients.   In order to substantiate
the theoretical predictions of pressure gradient effects,
purpose-designed rigs were used.   The experimental diffi-
culties were formidable because sealing of the assembly is
vital.   The conductivity values being measured are equivalent
to gas movements of a few meters per hour, and sealing significantly better than that must be achieved for the results
to be meaningful.   Gross and Scholz(29), and Naudin(23,30)
report similar German and French work.

3.   Standpipe tests.   The one insulation problem considered to be significantly different from those described
so far arises in the AGR standpipes.   In addition to the
thermal loading aspects, it is important to establish the
temperature of components and mechanisms within the standpipe.
The chosen solution was to build a full-scale test assembly
using prototype reactor components and a correctly-simulated
cooling system.   There are two basically different geometries
for the charge and control rod standpipes, and these were
mounted in turn on a large pressure vessel containing electrical heaters.   These two components therefore were tested
under full reactor conditions in every respect except that
the influence of pressure variations across the mouth of the
standpipe due to cross flows could not be established.

B. Material Testing

An extensive program of materials testing is required to
ensure that all materials used are compatible with the reactor gas environment and will survive without attention for
30 years. Many of these tests follow conventional practice,
and only the problems peculiar to insulation details will be
described in this section.

1. Fatigue. The critical component is the stud or strip
cover plate retaining feature. High strain fatigue data are
required to establish weld procedures and component life.

2. Resilience. It is necessary to carry out extensive
test work to establish that in the reactor lifetime, sufficient blanket pressure is maintained to ensure blanket support. To enable extrapolation of limited time test data (up
to 10,000 hours) obtained in isothermal test rigs to reactor
times and temperature gradient conditions, a mathematical
model has been developed(12).

3. Permeability. The basic function of the insulant is
to control gas movements; permeability measurements for isotropic insulants are required to establish the influence of
gas pressure, fibre diameter, blanket pressure and blanket
density.

The permeability of the insulation when installed on
the liner will be different in the normal and parallel directions, even when isotropic insulants are used, because of
sealing features. Large assemblies incorporating these features are used in tests to obtain the permeability data required for thermal performance predictions.

4. Fibre damage. Fibre insulants are contained by metal
components that vibrate as a result of the turbulence and
noise, and this movement must not cause fibre damage. Design
criteria are required. Small specimen shaker tests are described in Reference 12 and the data have been substantiated
in acoustically-driven diaphragm tests.

Exposure to reactor temperatures has shown that no
basic change of the fibre structure takes place and that the
fibre damage criteria are unchanged.

5. <u>Friction and wear</u>. In order to carry out routine inspection and maintenance work, the reactor is likely to be shut down annually, resulting in a full temperature cycle. Partial cycles are much more frequent, and result from power changes and unscheduled shutdowns. It is therefore necessary to establish that insulation components are free to accommodate differential movements. There are two major groups of components to consider:

    a.   massive sections such as coverplates that could create very large forces if constrained, but that will perform satisfactorily even if wear results.

    b.   thin sections such as foils that will buckle if loaded in compression and if damaged by cracking or wear could result in a deterioration of thermal performance.

C. Acoustic Testing

The power required to circulate the coolant in gas-cooled reactors is high (typically, 5-10% of the station electrical output) and some of this energy appears as noise. The noise levels vary around the circuit (Figures 2 and 3), being highest near the circulators; but in all areas, the insulation structure responds in a complex manner that can be determined only experimentally. A typical experimental assembly could consist of nine insulation modules in a 3 x 3 array, so the center module can be assumed to represent part of a very large surface. Such assemblies would be tested for response in a noise field representative of the reactor spectrum (typically broadband, with considerable energy at frequencies below 100 Hz). Corrections can be applied if the test spectrum does not match the reactor noise spectrum which, in the early stages of insulation design, is not likely to be known (31). Predictions therefore have to be made based on measurements taken in earlier designs of reactor vaults and circulators that have been subsequently checked in the commissioning phase. With this information and test data for noise spectra for the circulator design proposed, it has been possible to predict with fair accuracy the noise levels in the new design.

In the acoustic test program, it is necessary to

establish the effect of gas pressure on the structural response, and large, high-pressure test vessels are required
for these investigations. In general terms, it has been
found that the response, particularly at resonant frequencies, is reduced by increasing the gas pressure. Atmospheric tests tend,, therefore, to give pessimistic response
data. It is, however, much more difficult to achieve the
required noise levels at atmospheric pressure than at full
reactor conditions(12,32,33).

        Nitrogen is frequently used to model carbon dioxide
because its physical properties are more reliably known at
typical test pressures and temperatures. Acoustic tests are
required for all major insulation areas, and should always
include fault conditions such as stud failure.

D. Hot Vibration Tests

     Acoustic test rigs are expensive to operate, and large,
high-pressure facilities are limited in number. Furthermore,
it is not normally possible to reproduce the reactor acoustic
and thermal environments at the same time.

     A commonly-accepted procedure, therefore, is to measure
the response in acoustic tests under low-temperature isothermal conditions and then carry out endurance tests in the
correct temperature environment with the component movements
created by shakers driving the cover plates. Permeability
checks before and during the vibration tests are used to indicate any change in thermal performance(12).

     Hot vibration tests can be carried out most readily in
air, and wear data obtained thereby must be related to wear
in the correct reactor gas composition (Table I). With carbon dioxide cooled systems, this use of a modeling gas can
be justified. There is considerable doubt concerning the
truth of this for helium-cooled systems. If the tribology
problems only can be resolved in helium with typical trace
impurities, hot vibration testing is likely to become a major
cost in insulation development work.

E. Thermal Cycling Tests

     These tests are similar in many respects to the hot vi-
bration tests (IVD). The insulation specimens are, however,

now subjected to transient temperature conditions representative of complete and partial shutdown, load change, fuel
change and discharge, and other operational requirements.
Very thorough inspection of all insulation components is required before and after the test program so predictions can
be made of wear and deformation over the full 30 years of
reactor life.

V.   COMMISSIONING AND STATION PERFORMANCE

Early measurements taken before fuel has been loaded are
of considerable value in that design margins can be confirmed
while access to many critical areas is still possible.

A.   Acoustic Measurements and Response Tests

Running the circulators over full speed or flow conditions enables measurements of the acoustic field within the
reactor vault and of the response of insulation components
to be made.

B.   Flow Distribution and Pressure Gradient

Prior to the commissioning phase, all flow, turbulence
and pressure gradient data will have been based on tests carried out in scale models of the reactor upper and lower vaults
and special areas. Early confirmation of the predictions is
valuable, as the extrapolation of model data could lead to
error because of Reynolds number effects.

C.   Low-temperature Thermal Tests

Before fuel is loaded, the circulators or site boilers
can be used to raise the gas temperature so initial thermal
tests on the insulation at a uniform gas temperature (typically, 300°C for AGR) can be carried out. These tests reproduce only reactor operating conditions for areas such as the
floor and lower side walls, but using the measurements of
liner temperatures and heat loadings, predictions can be made
for all insulated areas at full operating temperatures. Faults
revealed by these preliminary tests include:

omission of sealing flanges(*36*)

failure of sealing details(36,37)

local gas flows bypassing the insulation
caused by circumferential pressure gradients
when operating with one or more shutdown
circulators(36,37)

unexpected sensitivity to severe local
pressure gradients(26,30)

inadequate sealing of penetrations(27)

inadequate cooling, requiring rearrangement
of penetration circuits so parallel operation
was substituted for series operation(27)

acoustic damage to foil insulation(38)

leaks in the cooling system (26)

D. Final Insulation Tests
Descriptions of the performance and faults of insulation
observed in reactor commissioning tests are given for:

Site	Reference
Oldbury	36, 37
EDF 3 and 4	34
Hinkley Point	14, 27, 39
Fort Saint Vrain	26

In general terms, the faults always have been with details
and these have resulted in high liner temperatures in small
areas.

The liner cooling systems have, in all reported cases,
been capable of rejecting the total heat loading(14,26,34).

E. Inspection Reports

In gas-cooled reactors, regular inspection of some of the
insulation is possible. This is particularly important near

the circulators where the potential for acoustic damage is
greatest. Oldbury has been operating for 10 years, and
statutory inspections have been carried out every two years.
Samples of insulation removed for laboratory examination
have shown some evidence of minor foil wear fretting and
carbon deposition(36), but liner temperatures have been
satisfactory. One reactor at Hinkley has been in operation
for two years, and detailed examination of the floor and
wall area in the boiler annulus, roof and gas baffle dome
has shown the insulation to be in excellent condition (14).

ACKNOWLEDGEMENTS

The authors wish to thank Mr. L. Broadley for his work
on the figures, and to the Directors for permission to pub-
lish.

REFERENCES

1.   Lord Hinton of Bankside. Conference on Prestressed
     Concrete Pressure Vessels. Institution of Civil Engin-
     eers, London, March, 1967, No. 751-2.

2.   Hannah, I. W., "Prestressed Concrete Pressure Vessels in
     the United Kingdom," Paper 121/75, pp 1-8. Experience
     in the Design, Construction and Operation of Prestressed
     Concrete Pressure Vessels and Containments for Nuclear
     Reactors, Institution Mechanical Engineers, York, England,
     September, 1975.

3.   Lamiral, G., Laurent, L., Bonnelle, R., Beaujoint, N.
     and Faurot, P., "The Prestressed Concrete Pressure Vessels
     for French Natural Uranium - Graphite - CO_2 Gas-Type
     Reactors." Third UN International Conference on the
     Peaceful Uses of Atomic Energy, A/CONF/28/P/52, May, 1964.

4.   Northrup, T. E. and Peinado, C., "Prestressed Concrete
     Reactor Pressure Vessel," Nuclear Engineering Inter-
     national, December, 1969.

5.   Turton, P. L., Hutchinson, L. and Old, R. A. B.,
     "Evaluation of Temperature and the Design of Cooling
     System for Pressure Vessels," Paper 61, 715-723 Con-
     ference on Prestressed Concrete Pressure Vessels,
     Institution of Civil Engineers, London, March, 1967.

6.   Barber, D., McLelland, G. S. and Seaton, A. R.,
     "Corrosion Protection of the Wylfa Vessel Cooling
     System," Paper 62, 725-733. Conference on Prestressed
     Concrete Pressure Vessels. Institution of Civil Engin-
     eers, London, March, 1967.

7.   Blanchard, G., Crutzen, S. and Farfaletti-Casali, F.,
     "Thermal Insulation System for Gas Reactor Concrete
     Pressure Vessels. Commission of the European Commun-
     ities Report EUR 5027e, 1973.

8.   Merot, J. P. and Lacroix, R., "Principle of 'Hot Wall'
     Insulation for Prestressed Concrete Reactor Vessels."
     Paper 177/75, 521-531, Experience in the Design, Con-
     struction and Operation of Prestressed Concrete Pressure
     Vessels and Containment for Nuclear Reactor, Institution
     Mechanical Engineers, York, England, September, 1975.

9.   Brockerhoff, P., and Scholz, F., "Problems of Thermal
     Insulation Systems in Gas-cooled Reactors." IAEA-SM-
     200/30, 353-362. Gas-cooled Reactors with Emphasis on
     Advanced Systems, Volume 1, International Atomic Energy
     Agency, 1976.

10. Freour, A. and Wurdig, P., "Protection Thermique par
     écrans a Circulation de Gas," Paper No. 34, 1035-1053.
     Second Information meeting on Prestressed Concrete Re-
     actor Pressure Vessels and their Thermal Isolation.
     EUR 4531, Brussels, November, 1969.

11. Wurdig, P., Freour, A. and Terpstra, J. S., "Thermal
     Insulation of Prestressed Concrete Pressure Vessels by
     'Gas Walls'," SM 111/10, 517-528. Proceedings of a
     Symposium, Advanced and High-temperature Gas-cooled
     Reactors, Julich, 21-25, October, 1968.

12. Furber, B. N., Hopkins, I. H. G. and Stuart, R. A., "The
     Development of Criteria for the Design of Insulation for
     Nuclear Reactors," Paper 133/75, 93-100. Experience in

the Design, Construction and Operation of Prestressed
Concrete Pressure Vessels and Containments for Nuclear
Reactors, Institution Mechanical Engineers, York,
England, September, 1975.

13. Hughes, J. W., Furber, B. N., Laing, G. W. and Armstrong,
    E., "Insulation Design and Development for the Oldbury
    Vessels," Paper 60, 703-713. Conference on Prestressed
    Concrete Pressure Vessels, Institution of Civil Engin-
    eers, London, March, 1967.

14. Furber, B. N., Colquhoun, J., Sheppard, M. and Owen, M.,
    "The Design Development and Commissioning of the AGR
    Insulation," Paper C120/77, 21-28. The Construction,
    Commissioning and Operation of Advanced Gas-cooled
    Reactors, Institution Mechanical Engineers, May, 1977.

15. Davidson, J., "The Heat Transfer of Metallic Foil Insu-
    lation in High-pressure Gas," Paper 13. High-pressure
    Gas as a Heat Transport Medium, Institution Mechanical
    Engineers Symposium, London, March, 1967.

16. Luikov, A. V., Shashkov, A. G., Vasiliev, L. L., Fraiman,
    Yu. E., "Thermal Conductivity of Porous Systems," Inter-
    national Journal Heat Mass Transfer, $\underline{11}$, 117-140, 1968.

17. Chang, S. C., Vachon, R. I., "A Technique for Predicting
    the Thermal Conductivity of Suspensions, Emulsions, and
    Porous Materials," International Journal Heat Mass Trans-
    fer, $\underline{13}$, 537-546, 1970.

18. Eckert, E. R. G., Carlson, W. O., "Natural Convection in
    an Air Layer Enclosed Between Two Vertical Plates with
    Different Temperatures," International Journal Heat Mass
    Transfer, $\underline{2}$, 106-120, 1961.

19. Jannot, M., Mazeas, C., "Etude Experimentale de la
    Convection Naturelle dans des Cellules Rectangulaires
    Verticales," International Journal Heat Mass Transfer,
    $\underline{16}$, 81-100, 1973.

20. Ozoe, H., Sayama, H., Churchill, S. W., "Natural Convec-
    tion in an Inclined Rectangular Channel at Various Aspect
    Ratios and Angles - Experimental Measurements," Inter-
    national Journal Heat Mass Transfer, $\underline{18}$, 1425-1431, 1975.

21. Furber, B. N., Davidson, J., "The Thermal Performance of Porous Insulants in a High-pressure Gas Environment," Paper 28. Second Meeting on Prestressed Concrete Pressure Vessels and Their Thermal Isolation, Brussels, November, 1969.

22. Viskanta, R., "Heat Transfer by Conduction and Radiation in Absorbing and Scattering Materials," Journal Heat Transfer, Trans ASME Series C, 87(1), 143-150, 1965.

23. Jannot, M., Naudin, P., Viannay, S., "Convection Mixte en Milieu Poreux," International Journal Heat Mass Transfer, 16, 395-410, 1973.

24. Micheau, P., "Exploitation des Resultats des Essais des Calorifuges Metalligues Cellulaires," Société Bertin et Cie Note Technique 68, C.g.10, 1968.

25. Menestrier, M. and Tarbes, B., Le Beton de Ponce Utilise Comme Isolant Thermique des Caissons en Beton Precontraint dans les Centrals Nucleaires E.D.F., 102-119. Nuclear Structural Engineering 2, 1965.

26. Jones, H. and Hedgecock, P. D., "Thermal Protection System for the Concrete Core Support Floor at Fort St. Vrain," Paper 135/75, 111-118. Experience in the Design, Construction and Operation of Prestressed Concrete Pressure Vessels and Containments for Nuclear Reactors, Institution Mechanical Engineers, York, England, September, 1975.

27. Colquhoun, J., Davidson, J. and Bolton, A. D., "The Design, Commissioning and Operation of Hinkley Point 'B' AGR Power Station Pressure Vessel Insulation," Paper 134/75, 101-109. Experience in the Design, Construction, and Operation of Prestressed Concrete Pressure Vessels and Containments for Nuclear Reactors, Institution Mechanical Engineers, York, England, September, 1975.

28. Smith, D. R., Bohm, E., Storrer, J. and Acciari, P., "Prospects for the HTR with Prismatic Fuel," SM-111/22, 181-195. Proceedings of a Symposium, "Advanced and High-temperature Gas-cooled Reactors," Julich 21-25, October, 1968.

29. Grosse, H. and Scholz, F., "Der Hochdruck Gaskanal Thieinig, Kerntechnik, 150-158, 1965.

30. Naudin, P., "Moyens D'essais de Calorifuges pour HTR,"
    Paper 31, 954-964. Second Information Meeting on Pre-
    stressed Concrete Reactor Pressure Vessels and Their
    Thermal Isolation, Brussels, EUR 4531, 1969.

31. Drake, M. E., "Gas Circulator Noise Generation Character-
    istics. A Comparison of the Effects of Helium, Carbon
    Dioxide and Nitrogen," Paper E S/7. Second International
    Conference on "Structural Mechanics in Reactor Tech-
    nology." Berlin, September, 1973.

32. Jones, E. A. C., "Acoustically Induced Vibration in
    Thermal Insulation Containment Structures for Advanced
    Gas-cooled Reactors," Paper 216, International Symposium
    Vibration Problems in Industry, Keswick, England, April,
    1973.

33. Jones, E. A. C., Hartley, W. and Hopkins, I. G. H.,
    "Response of Fibrous Prestressed Concrete Pressure Vessel
    Insulation Systems to Typical Noise Spectra," Paper H 5/6
    Second International Conference on "Structural Mechanics
    in Reactor Technology," Berlin, September, 1973.

34. Beaujoint, N. and Guery, A., "Experience in Operating
    and Inspection of Prestressed Concrete Pressure Vessels
    Belonging to Electricité De France," Paper 170/75, 457-
    464. Experience in the Design, Construction, and Oper-
    ation of Prestressed Concrete Pressure Vessels and Con-
    tainments for Nuclear Reactors, Institution Mechanical
    Engineers, York, England, September, 1975.

35. Irving, J., Smith, J. R., Eadie, D. M. D. and Hornby, I. W.,
    "Experience of In-service Surveillance and Monitoring of
    Prestressed Concrete Pressure Vessels for Nuclear Re-
    actors," Paper 169/75, 443-456. Experience in the De-
    sign, Construction, and Operation of Prestressed Concrete
    Pressure Vessels and Containments for Nuclear Reactors,
    Institution Mechanical Engineers, York, England, Septem-
    ber, 1975.

36. Brown, V. and Bland, A., "The Operator's View of the
    First Seven Years Service of the Concrete Pressure Vessels
    at Oldbury-on-Severn Power Station," Paper 168/75, 435-
    442. Experience in the Design, Construction, and Oper-
    ation of Prestressed Concrete Pressure Vessels and Con-
    tainments for Nuclear Reactors, Institution Mechanical
    Engineers, York, England, September, 1975.

37. Hughes, J. W. and O'Tallamhain, C., Contribution to the
    discussion, "High-pressure Gas as a Heat Transport
    Medium," Institution Mechanical Engineers, Symposium,
    London, March, 1967.

38. Williams, A. J., "Investigation into Structural Behavior
    of Insulation of the Prestressed Concrete Pressure Vessels
    of Wylfa Nuclear Power Station," Paper H6/5. First Con-
    ference on "Structural Mechanics in Reactor Technology,"
    Berlin, 1971.

39. McInerney, P. T. and Westerman, M., "Commissioning and
    Operation of Hinkley Point 'B' AGR," Paper C123/77, 47-
    55. The Construction, Commissioning and Operation of
    Advanced Gas-cooled Reactors, Institution Mechanical Eng-
    ineers, London, May, 1977.

NOMENCLATURE

Symbols	Description
B	permeability (Equations 2a, 2b, 2c)
C	inertial flow resistance term (see Equations 2a, 2b, 2c)
D	insulation pack thickness
F	dimensionless parameter (Equation 17)
$F_x(Ra)$	local convection contribution to the term k_x (Equation 5a et seq.)
H	insulation pack height
Nu	Nusselt number (see Equation 11)
$\overline{Nu}$	average value of Nu over the cold face of insulation pack
Q_{st}	heat leaving insulation pack cold face when no gas movement takes place
Q_{dy}	heat leaving insulation pack cold face when gas movement occurs
R	overall pack Rayleigh number (defined in Equation 14)

Ra	Rayleigh number for local convection (defined in Equation 6)
T	temperature
ΔT	temperature difference across an insulation pack
c	specific heat
d	distance between foils in foil insulation systems
g	acceleration due to gravity
k	thermal conductivity (tensor)
p	gas pressure
Δp	pressure drop across a seal (see Equation 10)
s	fibre diameter
v	(component of) the macroscopic velocity vector
x,y,z	Cartesian coordinates
α	see equations 5a, 5b, 5c et seq.
β	volumetric coefficient of expansion
γ	see Equations 5a et seq.
δ	see Equations 5a, 5b, 5c et seq.
ϵ	porosity of a porous medium
μ	viscosity
ρ	density

Subscripts

f	denotes properties of foil or fibre
g	denotes properties of gas
i	denotes properties of fibre insulation pack
s	denotes properties of seals
x,y,z	denotes properties of porous solids relative to the x,y,z directions

OUTAGE TRENDS IN LIGHT WATER REACTORS

E. T. Burns, R. R. Fullwood, and R. C. Erdmann

Science Applications, Inc.
5 Palo Alto Square, Suite 200
Palo Alto, California   94304

ABSTRACT

Operating experience in U. S. light water reactors
(LWRs) has shown that the impact of refueling outages on
plant unavailability is much higher than has been previously
anticipated. The purpose of this summary is to identify the
principal causes of the extensions of refueling outages, the
effect of these outages on plant productivity, and an alternate refueling cycle to reduce their impact. In order to
provide a perspective on the effects of refueling outages,
major outages greater than 100 hours in duration are also
included in this assessment. Both the refueling outages and
other major outages are displayed as a function of plant age;
this method allows identification of trends in these outages
as a plant matures. The result of the evaluation indicates
that utilities can improve plant availability by up to six
percent per year by increasing the time between refuelings
(i.e., from an annual to an eighteen-month refueling cycle).

I.   INTRODUCTION

Virtually all commercial nuclear power plants are base-loaded units; therefore, when a nuclear reactor is unavailable to produce power, replacement power must be provided to meet the electricity demand. In general, the replacement power will be generated by fossil-fueled units at a substantially higher fuel cost. Recent estimates of replacement power costs are on the order of $250,000(1) to $800,000(2) per day for a 1,000 MWe plant. In view of these high costs of alternate power generation, there is a real economic incentive to maintain high nuclear plant availability. Over the past few years, there have been a number of attempts to characterize the performance of nuclear plants(1,2); however, there has not been a detailed assessment of the causes of nuclear plant unavailability and the trends versus plant age. This paper presents a summary of:

a.   the principal causes of outages, ranking
      them by component

b.   the impact of these outages on plant
      productivity

c.   possible alternatives to improvement of
      plant availability

Refueling outages are found to be the major contributor to lost plant availability. The operating experience data for U. S. light water reactors (LWRs) shows that the length of "typical" refueling outages have been approximately 60 days, which is much longer than previous industry projections of 10(3) to 20(4) days. In addition to refueling outages, other outages that exceed 100 hours in duration also contribute a significant proportion of the lost plant availability. These two categories of outages combine to account for more than 80% of all plant outage time.

There is a significant effort within the industry to attain target plant availabilities of 86%(5) to 90%(6) for new plants. This paper identifies potential problem areas that may prevent the achievement of these goals based upon operating experience that indicates that current plant avail-

abilities are in the range of 73% for both PWRs and BWRs.
The data to support this analysis come from the approximately 250 reactor years of LWR commercial operation accumulated through June, 1977. By evaluating the trends in these
data as a function of plant age, the impact of refueling
outages is calculated.

The principal conclusions from this evaluation of nuclear plant outages are:

a.   Utilities can improve their LWR plant
     availability by approximately 6% if the
     refueling cycle is extended from an annual
     basis to an eighteen-month cycle.

b.   Plant unavailability due to component
     failures is dominated by a few major
     components (i.e., turbines, steam
     generators and pumps).

II.   IMPACT OF LONG DURATION OUTAGES

In this section, the magnitude of the potential improvement in plant availability will be investigated to provide a
"yardstick" on the losses attributed to various sources.
This section is important as a background setting for understanding the trends to be discussed later.

A.   Nuclear Plant Population

The population of nuclear plants considered in this report consists of 56 operating LWRs of diverse size and design.
Since each of the plants has been custom designed, caution
must be exercised in the use of the data. The best that can
be expected is that a characteristic trend can be identified
that will dominate the differences in design, construction
and size.

A profile of the nuclear plant population segregated by
BWR and PWR plants as shown in Figures 1 and 2, respectively,
may highlight differences in trends due to fundamental design differences. A comparison among the plants based on the
age of the units rather than on the calendar year of operation

is more useful since it isolates one key variable: plant
age. As will be seen in Section III B, there is an inherent
variation in equipment outages as a plant increases in age
from the initial "break-in" phase to a "mature" phase of
operation. While it can be shown that in some cases (e.g.,
snubber and pipe inspection) the calendar year of operation
is important in determining trends in the industry, it appears that the key parameter in isolating trends in major
outages is the age of a plant; therefore, plant age is used
here to display the plant populations. (In this report, the
age is measured from the date of initial commercial operation). Each of the plants of a given age included in the
population is treated equivalently despite the variations in
their size and design.

An important note in this profile of operating plants
(ages from zero to seven years) is that in some cases, the
initial seven years of data and the calendar period 1971-
1977 are mutually exclusive; therefore, plants such as
Yankee Rowe and Dresden 1, which do not have data in the
range of one to seven years during the period of 1971 to
1977 are excluded de facto. An added note of caution is
that the population contributing data in the period of five
to seven years of age is relatively small; therefore, outage
fractions from this portion of the analysis have a larger
uncertainty than in the initial four years where a larger
population provides greater statistical confidence.

While most systems of a typical plant conveniently follow the obvious division of plants into PWRs and BWRs, one
major piece of capital equipment does not lend itself to
this same classification: the turbine.

As has been pointed out previously(7,8), turbines represent a significant portion of the outage time associated
with nuclear plants. In addition, it has been pointed out(7)
that there has been a difference in the performance of turbines manufactured by the two major suppliers, General
Electric and Westinghouse. Since in the case of turbines,
the usual division of PWR versus BWR does not tell the entire
story, a separate population profile has been assembled for
GE and Westinghouse turbines. Figures 3 and 4 show the
population versus plant age. As a rule of thumb, it can be
said that all BWR plants contain GE turbines, while some
recent PWR plants have incorporated GE turbines also.

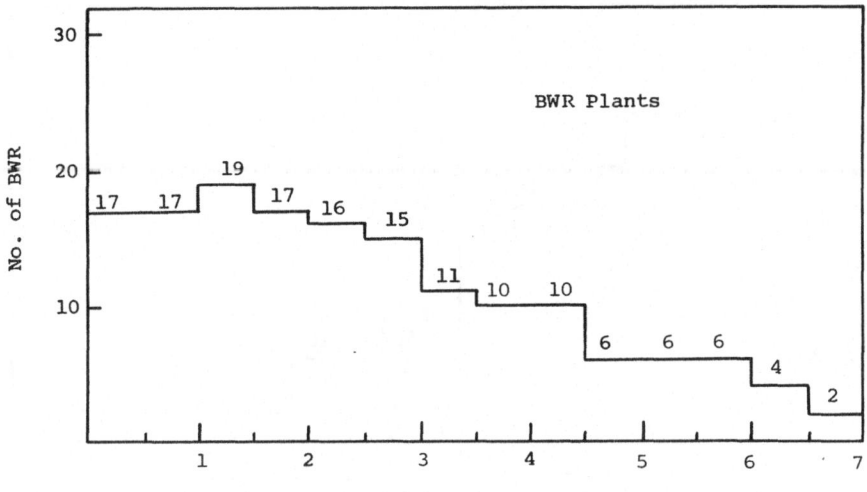

Figure 1.   Number of BWR Plants for which Data are
            Incorporated into this Report, Versus Plant Age

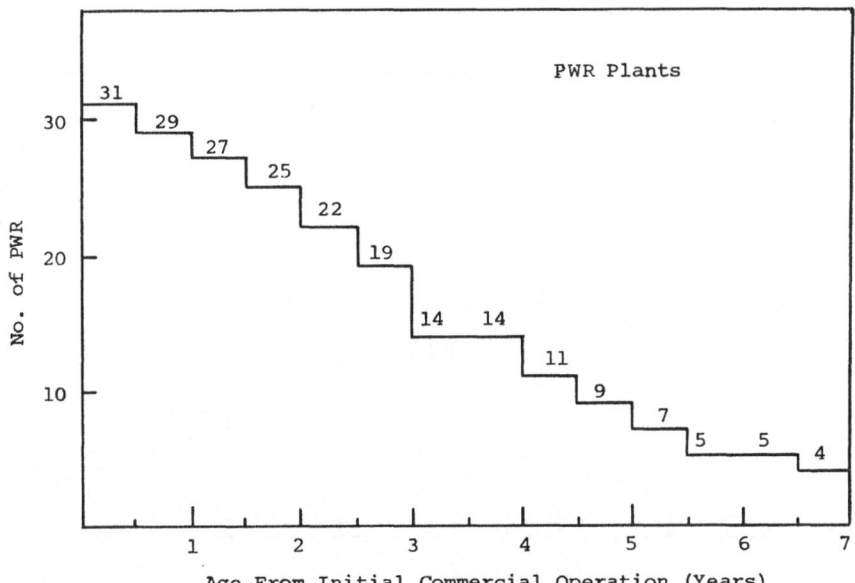

Figure 2.   Number of PWR Plants for which Data are
            Incorporated into this Report, Versus Plant Age

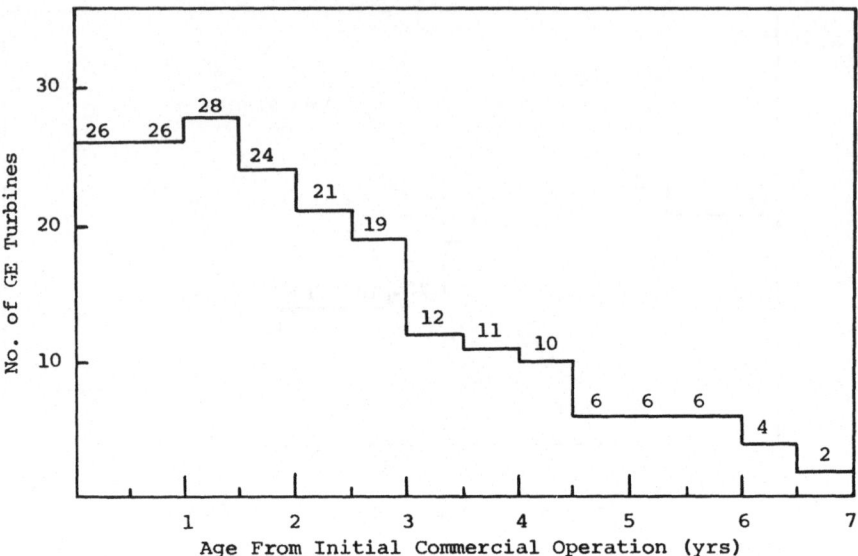

Figure 3. Number of GE Turbines with Data Included in this
Report, Versus Age from Initial Commercial Operation

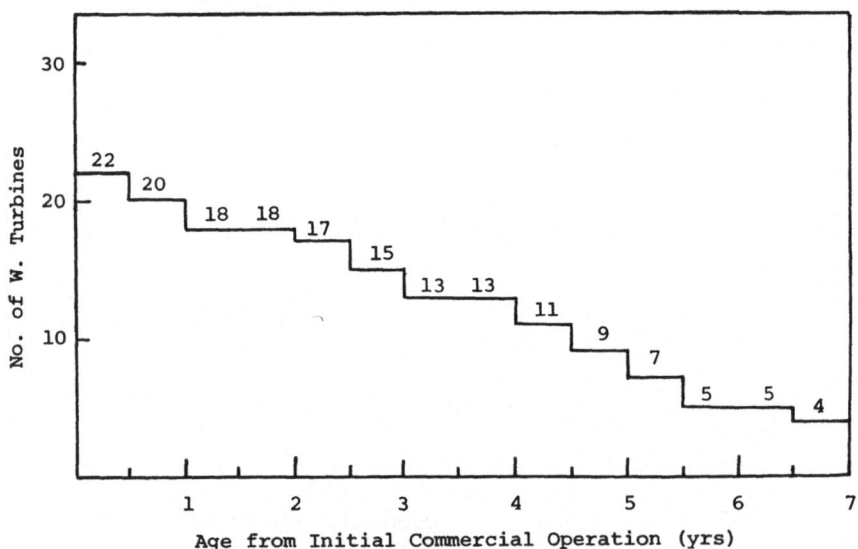

Figure 4. Number of W Turbines with Data Included in this
Report Versus Age from Initial Commercial Operation

B.   Profile of Nuclear Plant Performance

Having established the population of nuclear plants to be considered in this report in Section II A, this subsection summarizes the past power plant performance for this population. With this background, Section II C will then show the relative impact of major outages and refueling outages on plant performance.

There are several measures of nuclear plant productivity currently in use such as: plant availability, plant capacity factor (based upon maximum dependable capacity or design electrical rating), and forced outage rate. The capacity factor is the total amount of electricity actually produced by a unit in a year divided by the amount of electricity the unit could produce running at full capacity for the entire year.

Since the cost of nuclear generation of electricity is highly sensitive to plant availability and capacity factor, there has been a great deal of effort aimed at predicting trends in availability or capacity factor. However, there are two major problems in attempting to estimate future plant performance:

1.   There are virtually no data on large, mature units; i.e., those in the 1,000 MWe range that have been operating for several years. Estimates must be made based largely on experience with units that are smaller than those now being built and that are in their second through fifth or sixth years of operation. In addition, the capacity factors are a strong function of the plant electrical capacity. The capacity factors used in this report are based on the plant design ratings. This does not account for seasonal variations due to differences in cooling water temperature, or deratings due to environmental or safety considerations.

2.   Because of the diversity in plant design, size and age, the method of averaging plant performance parameters for these different units is not clear. One approach is to weight each unit in proportion to its

design rating. An alternative is to
weight all units equally regardless of
size. The latter method is used in
this section. A less defensible method
is to weight units according to the
energy they actually generate; however,
with this method, a unit that is not
operating (that has zero capacity factor)
simply drops out of the calculation.

Recognizing these limitations, this subsection seeks
only to crudely estimate the approximate magnitude of these
plant performance parameters. Therefore, for the purpose
of this summary profile, consider only those plants that
have completed at least one refueling cycle. Plant productivity can be summarized conveniently with a comparison of
availability or maximum dependable capacity (design) for
PWR and BWR plants. Figures 5 and 6 compare the cumulative
availability of PWR and BWR plants over their lifetimes. The
observations are not weighted; equal weight is given to observations from young, old, large and small plants. Note
that this aggregate comparison indicates that PWR and BWR
plants have approximately the same availability ($\sim$ 73%).
The main focus of this report will be on impact of refueling and major outages to plant unavailability; (it should be
noted that a similar comparison of capacity factor indicates
that BWR power restrictions have caused BWR plants to exhibit
a 6.6% lower capacity factor than PWRs).

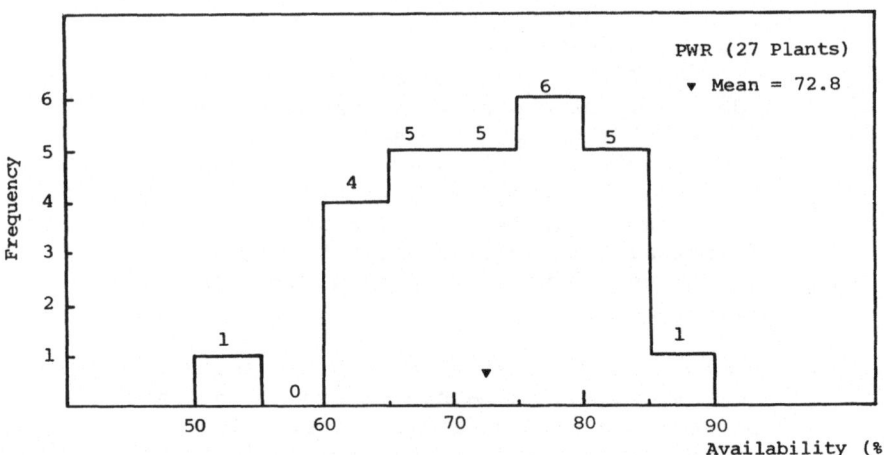

Figure 5. Frequency Histogram of PWR Plant Availability

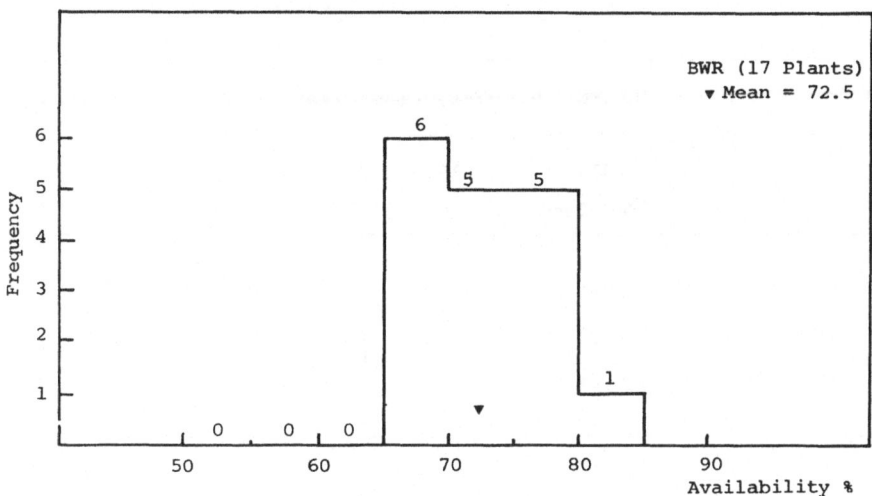

Figure 6.   Frequency Histogram of BWR Plant Availability

C.   Impact of Major Outages and Refueling Outages on
                    Plant Availability

        While major outages and refueling outages are intu-
itively judged to be important contributors to plant unavailability, the purpose of this section is to place the effects
of long outages in more quantitative terms.   Previous esti-
mates(9) of the effect of outages greater than 500 hours,
excluding refuelings, on a plant's capacity factor have been
in the range of 5%, based upon data through 1974.   The present
study has been expanded and updated to include: (1) outages
greater than 100 hours in length; and (2) data accumulated
through June, 1977.

        First, consider a gross comparison of the causes of
plant unavailability over the three-year period from May 1974
to June 1977.   Over this period of time, the fraction of una-
availability time attributed to various outages is given in
Table I.

TABLE I

Relative Contribution of Outages to Plant Unavailability
from 1974 through 1977

	PERCENT OF TOTAL OUTAGE TIME			
	1974 (May-Dec)	1975 (Jan-Dec)	1976 (Jan-Dec)	1977 (Jan-June)
Refueling	42%	32%	39%	51%
Outages >100 Hrs	39%	61%	32%	28%
Outages <100 Hrs	19%	7%	29%	21%

On the average, refueling contributed approximately
39.5% to unavailability over this period, and major outages
contributed 40%. The total plant unavailability time represents approximately 27% of the reduction in capacity factor.
Major outages contribute 40% to this, 11% to the reduction
in capacity factor; similarly, the refueling outages contribute approximately 11% to the reduction in capacity factor.

A graphical summary of the breakdown of plant unavailability is given in Figure 7. With this quantitative measure
of the impact of major outages and refueling outages on availability, let us now determine the systems and components that
are involved in major plant outages.

1. _Impact of Major OUtages on Plant Availability._ The
operating data are insufficient to determine the root causes
of the major outages; however, the identification of the components and/or systems involved in major outages will provide
additional information needed by utility and designer for
future decisions. First, the plant can be divided into the
nuclear steam supply system (NSSS) and the balance of plant
(BOP). While this division is somewhat arbitrary, and in
fact, differs from plant to plant, it gives a general overview of where the problem areas are located. The definition
used in this comparison is that all components inside the
main steam isolation valves are NSSS components. Using this

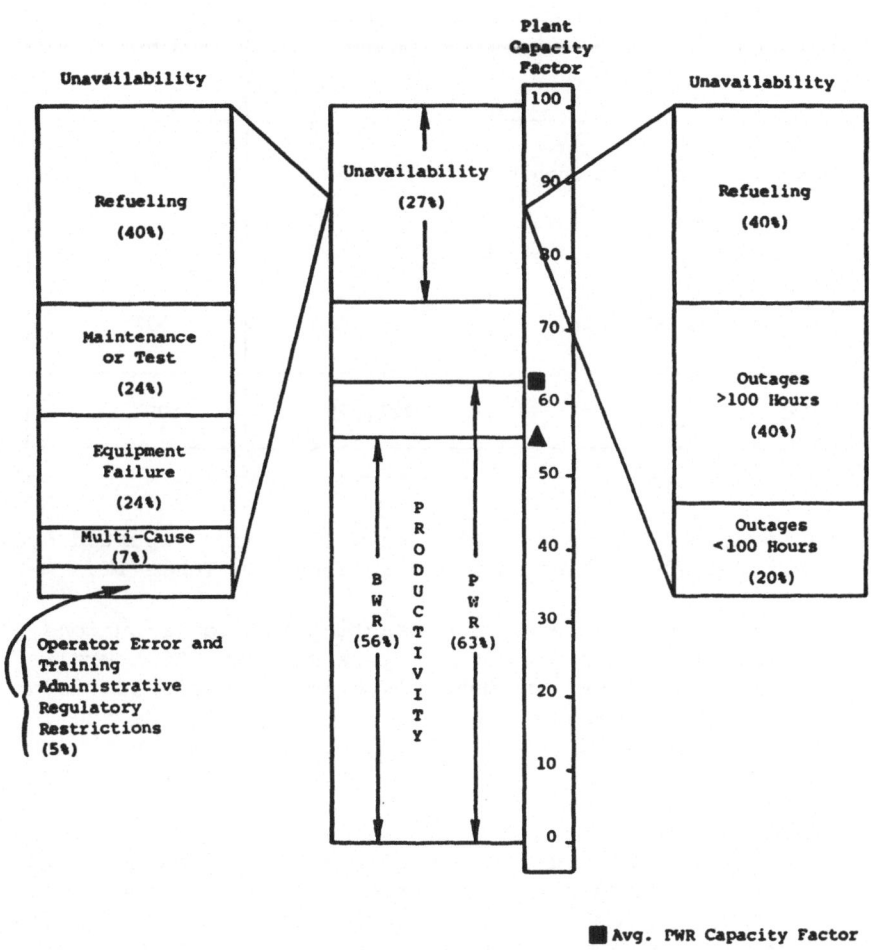

Figure 7.   Summary of Plant Performance, May 1974-June 1977

TABLE II

Comparison of All Outages >100 Hours in Duration
Over the Period 1971-1977

	Outage Duration (Hrs)	% of Totals
NSSS (Nuclear Steam Supply System)	103,358	47%
BOP (Balance of Plant)	118,501	53%
TOTAL	221,859	100%

convention, the outages greater than 100 hours reported over
the period 1971-1977 are divided as shown in Table II.

A more detailed way of dissecting the contributions
to unavailability due to outages greater than 100 hours is
to perform a breakdown by major "systems". Table III is a
ranking of those systems that have been shown to be the
cause of outages greater than 100 hours in duration by
operating experience over the time period January 1971 to
June 1977.

While the outages cited in Table III are associated
with the specific systems indicated, the outages contributing to these totals may be related to an equipment failure,
an inspection, a regulatory requirement, preventive maintenance, or some combination of these.

Table III is a composite sumary of 6 1/2 years of LWR
experience on 56 nuclear plants. It should be noted that
there is a wide diversity in the frequency of events of a
given type at each plant. For example, the reactor coolant
pump problems that have caused significant amounts of outage
and have primarily affected PWR plants, have varied from no

reported incidents of outages greater than 100 hours to a
large number of recurring problems over a period of years,
such as those experienced by Oconee 3 and Robinson 2. Similarly, condenser tube problems have plagued some plants,
such as Millstone 1 and 2, while other plants have encountered no major outages related to condenser problems. Each
of the systems in Table III is mutually exclusive except for
the pipe restraints/snubber system which is included as a
separate category since it has received a great deal of regulatory and utility attention at various times in the past.
However, many of the outages included under this category
also are included elsewhere because of coincidental work
being carried out during snubber inspection and repair. Also
note that the safety systems as used in the context of this
report include a wide variety of equipment whose sole purpose
is the safe operation of the reactor: containment, diesel
generators, high and low pressure injection systems.

Perhaps a more meaningful breakdown of these outages
would be a summary of the principal systems involved in
major outages for PWRs versus BWRs. Table IV points out the
sharp distinction in systems causing major outages in the
two reactor types. For BWR plants, the principal systems*
are reactor core-related problems, safety related systems,
and electrical systems (including the generator). None of
the top three contributors are the same for PWRs and BWRs,
indicating that there are significant differences in the
causes, and therefore, the remedies to BWR and PWR unavailability.

Since there are fewer BWR reactor years of experience,
Table IV provides only a relative ranking of the contribution
to major outages. A direct comparison of absolute magnitude
of the outage durations for systems can be made if the numbers are normalized to approximately the same operating
time (i.e., a normalizing factor of 1.4 times the BWR outage
times will yield a comparable base of comparison).

---

* Note that the outages due to fire are factored out of
this discussion since it is judged that a fire or
other rare event with high outage consequence also
could occur in a PWR.

TABLE III

Summary of Major Outages Compared by System

Rank	System	Total Outage Duration (Hrs) for Outages >100 Hours	% of Totals[*]
1	Steam System (includes turbine)	42,955	18.9%
2	Steam Generators	31,586	13.9%
3	Reactor Related[**]	29,994	13.3%
4	Fire[***]	26,360	11.7%
5	Reactor Coolant Pumps	21,443	9.5%
6	Electrical Systems (includes generator)	18,973	8.4%
7	Safety Related Systems	12,023	5.3%
8	Condensate System	11,765	5.2%
9	Feedwater Systems	11,270	5.0%
10	CDM Systems	8,312	3.7%
11	Pipe Restraints/Snubbers	7,462	3.3%
12	Off-Gas System	2,192	1.0%
13	Unknown/Not Specified	4,986	2.2%

[*]The total of outages >100 hours duration excluding refueling

[**]Includes BWR feedwater sparger and core spray pipe problems

[***]Does not include the six month outage at San Onofre due to a cable tray fire which occurred prior to 1971

TABLE IV

Summary of Major Outages by System for PWR and BWR Plants

BWR Plants		
Rank	System	Total Outage Duration (Hrs) For Major Outages
1	Fire	26,160
2	Reactor Related	20,599
3	Electrical	7,583
4	Safety Related	7,101
5	Steam	6,543
6	Recirculation Pumps	4,960
7	Condensate	3,903
8	Feedwater	3,632
9	Pipe Restraint	2,526
10	Off-Gas	2,192
PWR Plants		
1	Steam	38,366
2	Steam Generators	31,375
3	Reactor Coolant Pumps	16,483
4	Reactor Related	9,395
5	Condensate	7,862
6	Feedwater	7,638
7	Safety Related	4,922
8	Pipe Restraints	4,605
9	Electrical	4,411

2.   <u>Lost Plant Availability Due to Refueling Outages.</u>
The character of refueling outages generally is quite different from that encountered in other major outages. Most
major outages occur with very little advance warning; therefore, there is a minimum amount of preparation involved.
On the other hand, refueling outages tend to involve more
extensive planning, but they also have a great deal more
associated work occurring in parallel with the primary task of
refueling the core. If a critical path analysis is performed
on a selected group of refuelings, those operations that are
the largest contributors to "typical" refueling outages can
be identified. Such a summary review may identify those
areas of this exceedingly complex outage, called a refueling, that are most fruitful to pursue for potential decreases
in the overall refueling length. An overview of these data
can be obtained by considering a comparison of the average
PWR and BWR refuelings that are summarized in Figure 8.

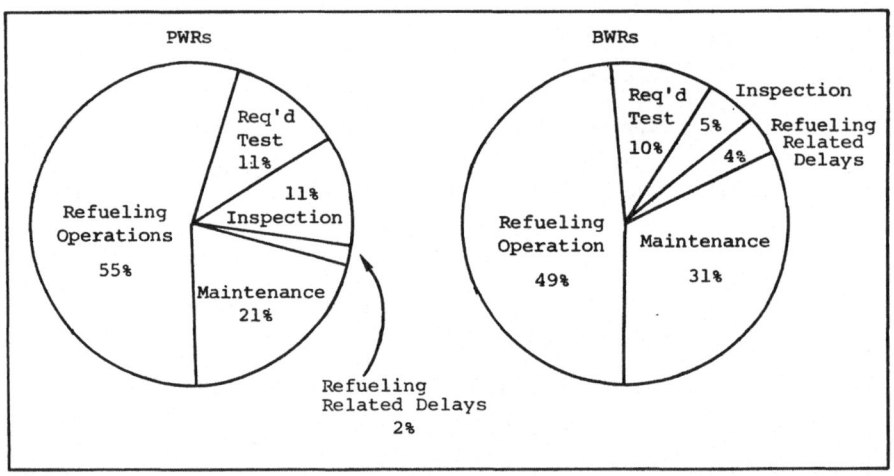

Figure 8.   Graphical Summary of Typical Refueling Outages
            Based Upon a Selected Sample of Refuelings

It should be carefully noted that since very long refuelings (i.e., refueling outages >4 months) have not been
included in the group of selected refuelings, Table IV is
biased in that the overall percentage of maintenance may be
underestimated. However, since we are most interested in
the "typical" refueling, it is judged that Figure 8 represents an accurate summary of the events that comprise typical refuelings. Based upon the simplistic breakdown of a
typical outage in Figure 8, approximately one-half the
critical path time during a refueling outage is associated
with operations not specifically required to refuel the
plant such as: (a) maintenance (20%-30%); (b) required
testing or inspections (15%-20%); and (c) refueling-related
delays (~4%). The other half of the outage time is associated with the actual refueling operation.

     While there is always a large number of maintenance
and repair items that are performed during refuelings, utilities generally attempt to perform the maintenance during non-
critical path time. However, there are some instances when
the repair may interfere with the critical path work (e.g.,
in-core repairs, steam generator tube plugging in some plants)
or may extend beyond the minimum refueling time (e.g., turbine repair). For example, there have been much longer refuelings than considered in this section that have been
totally controlled by maintenance or repairs (e.g., the 1973
Haddam Neck refueling that lasted approximately 160 days
required replacement of the turbine rotors). Nevertheless,
attempts to dissect the contributing causes of refueling extension due to specific types of maintenance would not be
precise if based upon the limited data sample available because of the wide variation in the types of maintenance that
can lead to extended outages.

     Even the same maintenance operation at two plants can
have significantly different impacts on the length of the
outage. One example of this variability in impact of maintenance operations is in the case of steam generator tube
repair. Westinghouse PWR plants with three loops have no
loop isolation valves, which means that in their present
configuration, steam generator (S/G) tube plugging or eddy
current testing must be performed with the water level below that of normal refueling (i.e., below the reactor vessel
nozzles). This means that a portion of the S/G work must be
performed in critical path time, usually in series with head

removal, therefore having a direct impact on plant availability. B & W recently has developed a technique that may
reduce critical path time required for eddy current testing;
the technique allows eddy current testing to be performed
"wet" (34).

The largest segment of a typical refueling outage
(~50%) is directly related to operations necessary to refuel the reactor; however, in pursuing methods of reducing
the total outage time, it is necessary to know how this
effort is apportioned among the various operations. Figure
9 provides the details of the length of time required for
various operations for the selected refuelings. The comparison between PWRs and BWRs highlights the differences in
refueling approaches in the two types of plants.

Because of the variations in PWR and BWR plant design,
there can be substantial differences in the time required to
perform similar tasks in PWRs and BWRs. From Figure 9 it
appears that PWR closure head removal and reassembly operations take longer than comparable BWR operations. However,
in the area of fuel movement, the situation is reversed, and
PWRs have a distinct advantage over BWRs, based upon the
operating histories from the selected sample of refueling
outages.

Possibly the most dramatic fact to be noted from Figure 9 is the large percentage of refueling outage time associated with testing and inspection. The combination of these
two categories accounts for approximately 20% of the "typical"
refueling outage. In terms of plant capacity factor, this
outage time represents nearly 4% of the total energy that
can be produced in an entire year, which translates into
approximately 12-14 days. It is important that utilities,
component vendors and regulatory agencies are aware of the
contribution of testing and inspection to power plant operation. Alternate testing methods, less frequent testing,
and on-line testing are items that should be factored into
the conceptual plant design and arrangement. Layering
additional test requirements on the utility can hamper its
ability to operate efficiently with the end result being
lost productivity. Safety cannot be relaxed or compromised.
However, a question that does deserve consideration is,
"What is the optimum balance between testing and power
production?"

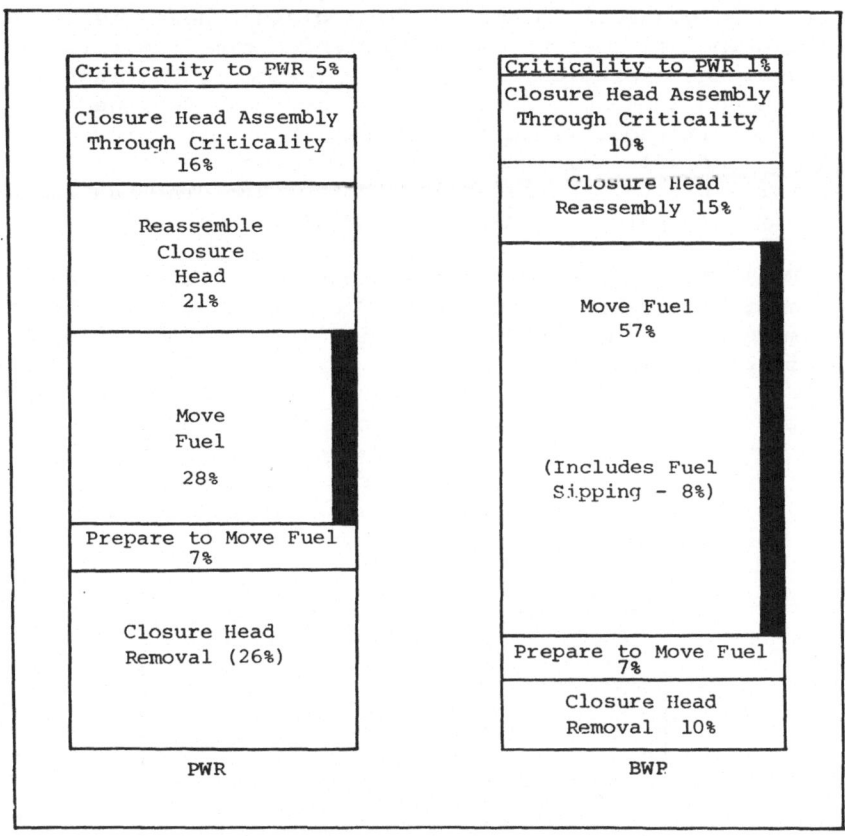

Figure 9.   Comparison of Refueling Operations for
Selected PWR and BWR Plants

        Closure head removal and reassembly in PWR plants takes
longer than in BWRs, principally due to the amount of associated equipment that must be handled in PWR plants. Specifically, in PWR plants, the control rod drive mechanisms
(CRDMS) penetrate the top closure head, while in BWR plants,
the CRDMs penetrate the bottom closure head; therefore, in
PWR plants, the movement of fuel requires disconnecting all
the CRDMs from their cabling and cooling piping in order to
remove the closure head. In addition, because of the higher
primary system pressure in PWR plants, the higher closure head
head preload requires additional time to detension and tension the closure studs. The disassembly and assembly of the

closure head connections and the high closure head preload
are the principal reasons for the extended time for this
portion of the operation. For future reactors, Westinghouse
has developed a closure head assembly schedule designed to
reduce the time to perform this operation by utilizing a
unitized head area arrangement that contains the studs as
an integral part of the assembly, thereby avoiding the
necessity of separate handling of the closure studs.

Fuel movement in BWRs takes appreciably longer than
the comparable operation in PWRs. The reasons for the longer
"fuel movement" times in BWRs principally are due to a combination of the following:

a. There are approximately four times the number
   of fuel assemblies in BWRs, compared with
   PWRs for the same rated power. For example,
   consider the following comparison of two 800
   MWe units:

Plant	No. Fuel Assemblies
BWR	720
PWR	157

b. BWRs sometimes reuse fuel channels. This
   requires some coordinate effort, and therefore, potential delays in interchanging
   fuel channels during the fuel movement
   operation.

c. Leaking fuel has caused a larger percentage
   of operational problems in BWR plants than
   in PWRs. More "leakers" in a core leads to
   higher coolant activity and off-gas releases.
   Since this may in turn lead to power deratings
   to limit the off-gas activity, it is prudent
   to remove the leaking fuel assemblies from
   the core. The best opportunity for removal
   of the fuel is during a refueling. However,

since only part of the core is removed
at each refueling, identification of
the fuel assemblies containing leaking
fuel rods becomes a necessary operation.
Identification of leaking assemblies is
performed by fuel assembly sipping.
This may be performed in or out-of-core
for BWRs and only out-of-core for PWRs.
The sipping process can add significantly
to the critical path time in BWRs, either by:

1. requiring controlling time to sip
    fuel assemblies in-core;

2. requiring controlling time to re-
    move all fuel assemblies to allow
    out-of-core sipping.

PWRs have not had a chronic problem with
leaking fuel, and therefore, all fuel
assemblies generally are not sipped.
Instead, only those assemblies removed
from the core are sipped, and sipping is
performed in non-critical path time.

The above summary of operations that occur during a
typical refueling outage provides a perspective on the relative distribution of plant unavailability due to the different types of refueling operations and the associated maintenance and testing.

III.   TRENDS IN REFUELING OUTAGES AND MAJOR OUTAGES IN LWRS

The amount of LWR operating experience has increased
quite rapidly in recent years; the number of years of operating experience increased by approximately 30% in 1975, and
an additional 28% in 1976. Therefore, it appears fruitful
at this point to determine whether a trend as a function of
maturing plants can be isolated in the complex operations of
a nuclear power plant. If a trend can be established, one
of the key elements in the planning sequence will be achieved:
the identification of the problem, its magnitude, and its
anticipated variation with time. As is evident from a re-

view of the data, there is a wide diversity in the types of
outages occurring; however, there are certain classes of
high impact outages that can be isolated, and to a large
extent, either prevented from occurring or adequate preparation made to minimize impact on plant performance. A known
trend at one plant can aid management decisions at other
utilities in the preparation of procedures, identification
of sources of spare parts, and the training of personnel.

A.   Overall Outage Trends as a Function of Plant Age;
      Comparison of PWR Versus BWR Plants

     A continuing theme in the literature has been the call
from both industry and utility management personnel(10,11,12,
13,14) for additional planning and preparation for major
outages, including collection of data and the application
to special maintenance tasks. It is recognized that in order
to minimize forced outage time, a comprehensive program of
preventive maintenance must be incorporated into a work
package. To establish a successful preventive maintenance
program, utility management and engineering personnel must
be informed of the outage trends of similar equipment throughout the industry. This section takes a broad-brush look at
the overall trends in long duration outages. More specific
information that may be required by utility planners and
vendor design engineers to identify trends in specific types
of equipment is provided in Sections III B and III C. Of
course, this type information can only alert the utility
planner and designer to potential problem areas. For specific failure mechanisms and times to failure, additional
information is required that only is available through the
collection, evaluation and sorting of detailed operating
experience data currently not available.

     The LWR operating experience outage data for the six
and-a-half year period 1971-1977 includes 480 major outages
and 128 refueling outages. The number of plant years in
each analyzed category is calculated from the data in Section II A. Using this as the population base and using the
major outage hours, the average outage time per plant year
can be calculated. This is the parameter that is used in
the remainder of the report to characterize potential trends.

     Figure 10 summarizes the overall outage trend of U. S.
LWRs during their initial seven years of commercial operation

on a per plant year basis. The trend confirms a portion of
previous predictions from the electric power industry(4)
that as nuclear power plants mature, their productivity
(availability) will increase. Clearly, however, this trend
in major outages is only a segment of the plant productivity
picture, which also includes refueling outages, short duration outages, and power restrictions. In the case of the
major outage trends, there is a distinctive maturation trend
for LWRs. After the second year of commercial operation,
the major outage time required on a per plant basis is dram-
atically decreased to an approximately constant level. This
indicates that after a high initial outage rate due to
special "start-up" or "break-in" problems, nuclear plants
settle into a constant background level of major outages.
It must be understood in interpreting these trends that the
calculated outage rates for the sixth and seventh years of
operation have greater uncertainty bands on the calculations
since the number of data points (plants operating) is substantially less than during the initial years.

In order to emphasize the fact that major outages represent only one contributor to reduced plant availability,
note that during the first year of commercial operation,
generally less than 5% of the plants have a refueling outage.
In subsequent years, 80%-90% of the plants undergo one refueling each year. Since refueling outages historically
have taken an average of 2.3 months, they represent a significant impact upon plant availability. Therefore, an accurate representation of trends in overall plant availability
must account for both of these effects. Figure 11 shows that
if refueling outages are included, the first two years of
operation still represent the years with the highest outage
rate per plant.

Nothwithstanding this caution in the use of the major
outage trends, the distribution of major outages represents
one important contributor to the understanding of LWR performance. The understanding of each contributing cause of
plant unavailability will lead to better planning for load requirements, improved maintenance preparation and possible
changes in the equipment design or arrangement.

An important trend in Figure 11 that needs to be emphasized is that after a plant has reached maturity (i.e., after
the second year of commercial operation), refueling outages
tend to account for more than 40% of the plant unavailability

time which was estimated in Section II, where the estimate
was based upon a plant sample biased toward young plants.
By unfolding the refueling outage information presented in
Section II according to plant age, a pattern in the contribution of refueling outages emerges. The trend from Figure
11 emphasizes the fact that as plants mature, refueling outages will take on continually increasing importance in determining overall plant availability. This trend suggests
that the total effect of refueling outages on plant unavailability during equilibrium fuel cycles approaches 60% of all
the unavailability time in years after the second year of
commercial operation.

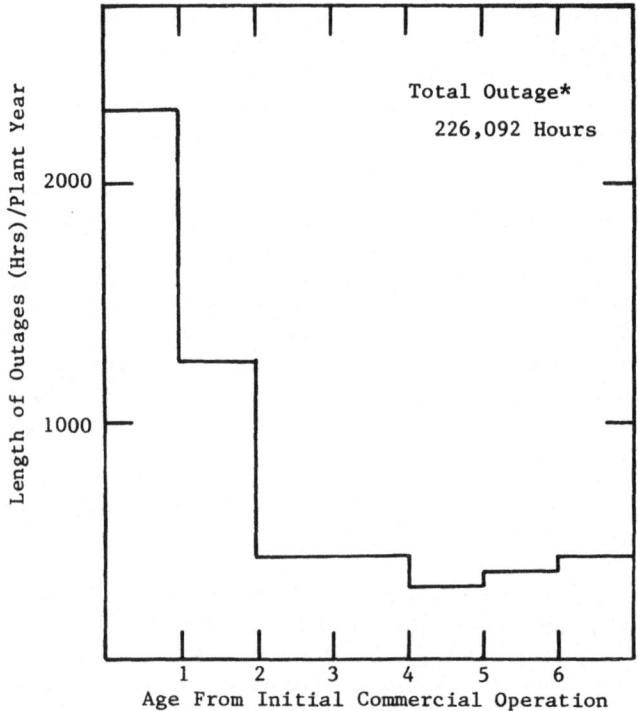

Figure 10.   Trend in Major Outages* for LWRs

* Only non-refueling outages >100 hours are included here.

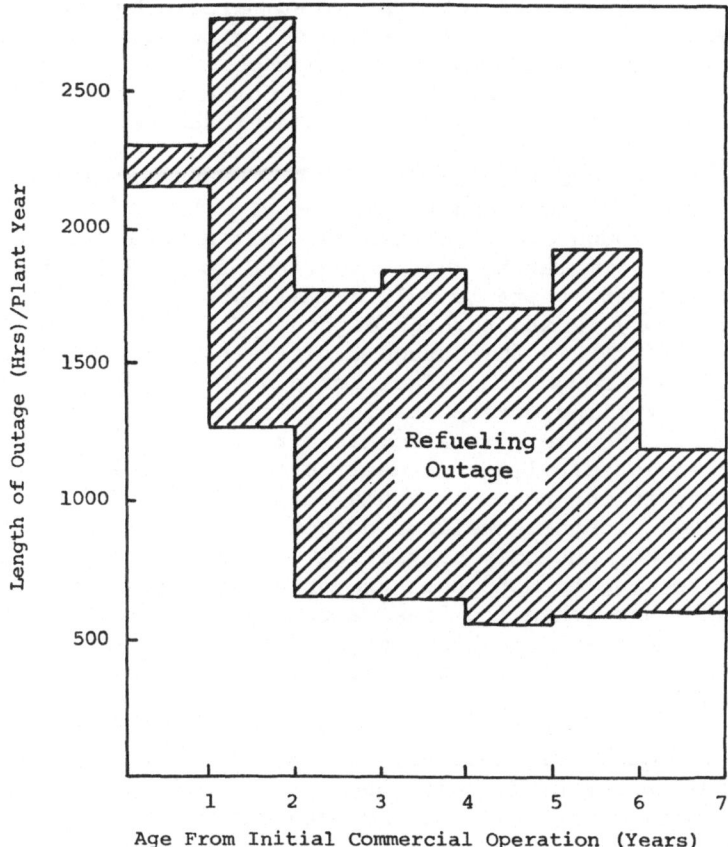

Figure 11.   Distribution of LWR Outages as a Function of
Plant Age

        If refuelings indeed were annual affairs, as is often
suggested in the literature, Figures 10 and 11 would portray
all the information needed to assess the trends of refueling outages. However, there is actually a wide distribution
in both the refueling outage duration and the intervals between refuelings. Therefore, we present here a comparison
of the refueling outage length as a function of the refueling cycle number. Figure 12 gives a composite frequency
histogram of the number of refueling outages as a function

of outage length for all refuelings, summarized by fuel
cycle number. This composite of refueling outages is a key
to the understanding of typical nuclear power plant performance trends for mature plants. The important item to note
from this comparison is that it shows a pronounced difference between the first refueling outage (12.9 weeks or ~3
months, and subsequent refuelings of 9.9 weeks or ~2.3
months). The one exception to this conclusion is the combined sixth through eighth refuelings that have a mean refueling length similar to that calculated for the first refueling. However, there is only a small population of plants
in this category (sixth through eighth refuelings), and there
is a large scatter in the outage times. In fact, the population of 6th through 8th refuelings has the highest percentage of refueling outages taking less than six weeks of
any category. However, it also has a relatively large number of extended refueling outages. Because of the large
uncertainty, it is judged that the combined data for the
sixth through eighth refuelings do not represent the beginning of a trend, but only statistical scatter in the data.
It is interesting to note that the group of refuelings that
historically has taken the least amount of time is the second
refueling; its mean refueling outage time is two months,
which still is much longer than previously expected by reactor suppliers.

    In subsequent discussions, the focus of attention will
be on mature plant performance and on the dominant trends
affecting plant availability. As previously discussed, the
"first-refueling" population represents a disproportionate
fraction of the total population and could bias the results
of any evaluation that did not account for this fact. Therefore, the mean refueling outage time considered in other sections will be 9.9 weeks (~2.3 months). In particular, the
mean refueling outage time of 9.9 weeks will be used in
Section IV to discuss the advantages and disadvantages of
extending the refueling cycle to 18 months in order to improve plant availability.

    Thus far, we have not discussed the scatter in these
data. For each refueling, first through eighth, there is
at least one case of a refueling being accomplished in a
"short" time (i.e., four to six weeks). While four to six
weeks has not been considered a short refueling in the past,
the operating data indicate that this is the fastest time

that can be expected under today's design and testing conditions. On the other end of the scale, there are a number of extremely long refueling outages whose length has been determined by plant problems that were corrected during the outage (e.g., feed-water sparger inspection/replacement, core vibration repair, turbine repair). However, it is apparent from the data that while long refueling extensions are infrequent, nevertheless they continue to occur in the population, and therefore, must be considered in the evaluation of plant availability. For the moment, if we neglect all refueling outages greater than 12 weeks as not relevant to the discussion of typical refueling outages, then the average refueling outage is reduced from 2.3 months to slightly less than two months (which is still much higher than has been expected in the past).

An important facet of the scatter of the data is the variability from plant to plant. In terms of individual plants, there are some plants that have been able to consistently achieve short refueling outages (e.g., Robinson 2, Point Beach, Vermont Yankee and Haddam Neck), and there are some plants that are on a longer learning curve and have not consistently refueled their units within the determined mean time.

While our attention primarily is oriented toward the United States light water reactor population, it is useful to measure the way this population compares with other, similar units in the world. The population chosen for comparison is that of the European light water reactors. Sharply different concepts in reactor designs (e.g., gas-cooled reactors and heavy water reactors) are eliminated. Hopefully, this simple comparison will provide a degree of insight into the state of technology, planning, maintenance and approach to the refueling outage. The principal aim of this section is to obtain an appreciation of the comparative lengths of refuelings when different philosophies in maintenance and regulatory requirements are applied.

Some qualifiers on the European LWR data are used here for purpose of comparison: the European plants, in general, are smaller than the plants in the United States population, and data concerning the European plants are available only through 1975; therefore, the data have at least two bias factors built into them that may distort the results.

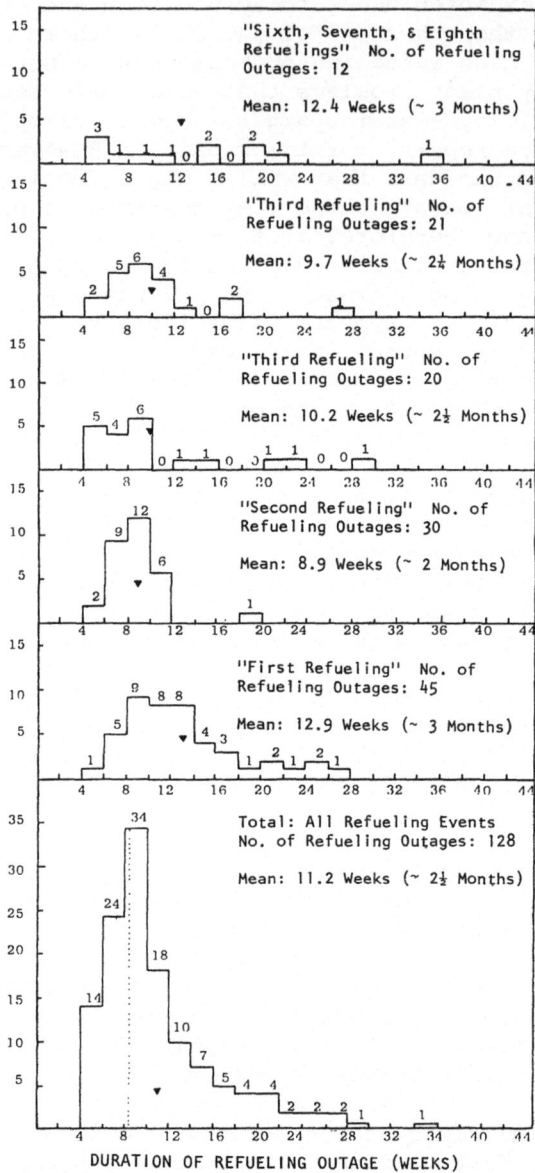

Figure 12.   Frequency of Refueling Outages as a Function of
the Length of the Outage; Composite by Number
of Refueling, First Through Eighth

Since there is approximately the same percentage of
first refuelings in the European population (26%) as in the
United States population (35.1%), it is appropriate to consider the distribution of outages based upon the entire population of refuelings. The comparison of the distribution
of refueling outages by length for United States LWRs versus
that for European LWRs (Figure 13) shows an interesting fact:
while European units incur a small number of long duration
refueling outages as U. S. plants do, they have achieved a
significantly lower median refueling duration (i.e., seven
weeks versus nine weeks for U. S. reactors). From Figure 13
it is clear that a large percentage of the European refuelings are accomplished in a very short time (in the four to
six week time frame). Based upon the comparison in Figure
13, it appears that, in general, the LWRs operating in
Europe experience shorter refueling outages than those in
the United States.

B.   Trends in Major Outages in LWRs

To determine whether there are any distinctive trends
for PWR and BWR outages, the overall trend of major outages
from Figure 10 is divided into trends for PWR and BWR plants.
Figure 14 shows the variation in the average outage time per
plant year from initial commercial operation through seven
years. BWRs and PWRs both exhibit the same tendency for
high outage rates on a per plant basis during the initial
two years of plant operation. Previously, however, it was
pointed out that the causes of this high initial outage rate
are different for PWRs and BWRs related to differences in
systems and design philosophy.

The trend of outages as a function of plant age can be
dissected into individual component contributors. This
technique will highlight any differences in the impact of
specific components on plant availability. Using the operating experience data cited in Section II, consider the top
four component contributors to major outages as shown in
Table V. These four components (i.e., turbine generators,
steam generators, pumps and core internals) are discussed
in more detail in the following summary.

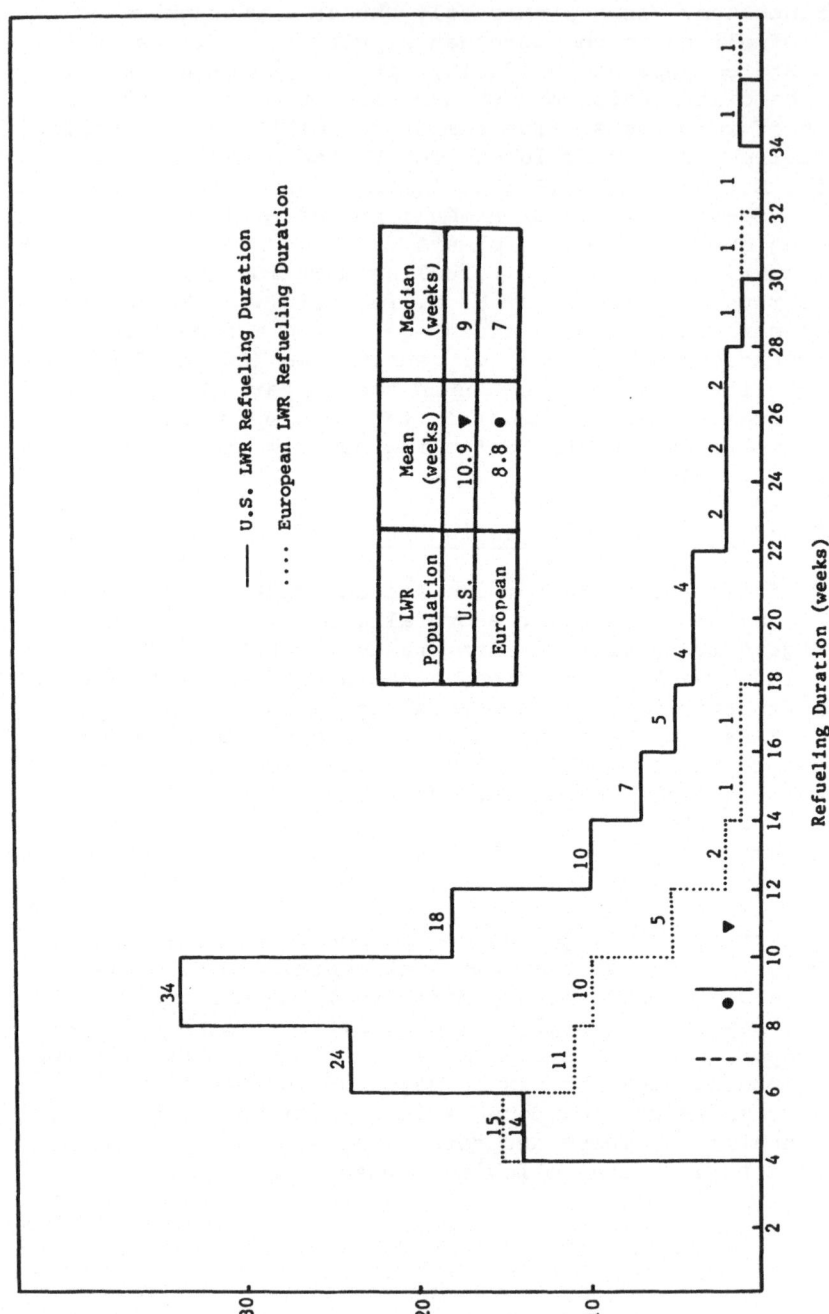

Figure 13.   Refueling Duration: Comparison Between United States and European LWRs

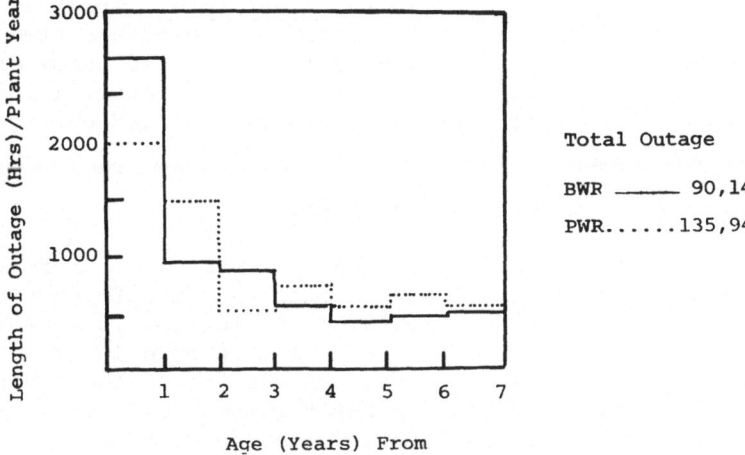

Figure 14.   Average Yearly Major Outage Time Per Plant
As a Function of Plant Age

TABLE V

Summary of Major Outages of Component Type

Component Types	% of Total
Turbine-Generators	18.5%
Steam Generators	13.9%
Pumps	9.9%
Reactor Core	9.8%
Valves	8.1%
Pipe*	7.1%
Condensers	5.2%

*Includes BWR core spray, recircula-

tion, and feedwater sparger

1.   Outage Trends in Turbine Generators. Turbine-
generators have caused a major portion of the outage time
occuring outside of refuelings at nuclear plants. From Figure 15, note that steam turbine failures have caused a large
percentage of the major outages that have occurred during
the first year of commercial operation; also, there is an
apparent trend of increasing turbine-related problems in the
sixth and seventh years of plant operation.

Previous studies(7,8) have identified a difference
in the performance of turbine-generator systems of the two
major suppliers. In view of this, a distinction is made in
this report between the turbine vendors. Figure 15 shows
the trend in outages as a function of the age of the plant.
The data show that Westinghouse LWR turbines seem particularly troubled by blade failures, while the GE turbine-
generator problems primarily have been generator related.
Westinghouse units have incurred nearly six times the outage
time for GE units, based both on a total outage time and on
a per-plant-year basis.

A summary of turbine blade failures in LWRs is
presented in Figure 16 which is a time line displaying the
plants and dates for the blade failures affecting plant availability (note that all units' have Westinghouse turbines).

Figure 16 indicates that major turbine blade failures are distributed throughout the six and one-half year
period included in this study; however, Figure 15 shows that
the bulk of the outage time exclusive of refueling extensions
occurs during the first year of commercial operation of most
plants. The trend in the outage frequency per plant shows a
significant drop in the second through the fifth years. Although the turbine-generator outages still make a significant
contribution to the overall outage time after the second year
of operation, the decreasing trend with plant age indicates
that the initial problems primarily are design related, and
surface early in plant life. However, among the small plant
population in the sixth and seventh years of operation, there
have been some blade failures resulting in lengthy outages.
This incidence of failures corresponds to the five-year cycle
of operation recommended by the turbine vendor before complete
overhaul is recommended. As operating experience begins to
accumulate for operation beyond five years, turbine failures
should be continually monitored to determine if a trend is

developing that may affect LWR plant productivity adversely as plants age.

2.   Outage Trends in Steam Generators. The summary of outages in Table V indicates that steam generators (primarily tube failures) are major contributors to plant unavailability, causing 13.9% of the major outages in LWRs, or 23% of the major outages in PWRs. The causes of the outages are, for the most part, related to a gradual deterioration of tubes over time; however, the outages also can be classified according to specific actions that are required to:

a. Plug Tubes: To prevent leakage of radioactive primary coolant into the secondary system, failed tubes or those with incipient failure are plugged.

b. Inspect Tubes: To determine the integrity of tubes and to chart its variation with time, an Eddy Current Test is performed on steam generator tubes.

c. Change Secondary Chemistry: To reduce corrosive tube attack, secondary water chemistry has been changed from phosphate to all volatile chemistry treatment (AVT).

d. Remove Sludge: To reduce the chloride stress corrosion cracking that is aided by sludge accumulation, efforts are made to remove the sludge by flushing the secondary side of the steam generator.

The steam generator tubes are thin-walled ($\sim$1-2mm) members that were expected to last the life of a plant with a small number of failures.*

----

*   Each steam generator is fabricated with a small percentage of excess capacity (i.e., larger number of heat transfer tubes than required) in anticipation of minor problems necessitating plugging of a small number of tubes due to fabrication defects of accelerated corrosion.

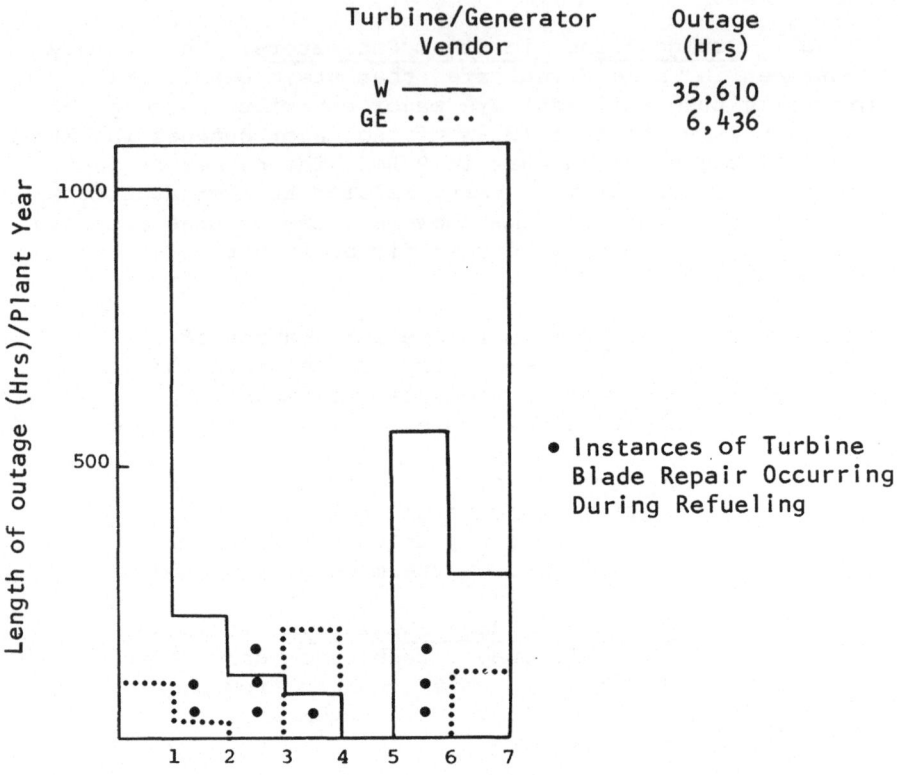

Figure 15.   Trend in Average Major Outage Time Per Plant
             Attributed to Turbine-Generator Components
             (excluding Refueling Extensions)

However, experience to date indicates that these thin members are susceptible to a wide variety of failures. A summary of the dominant mechanisms of tube failure in LWRs includes the following(15,16,17,18,19):

    a.   Stress corrosion cracking of both
         stainless steel tubes and inconel
         tubes has been observed.

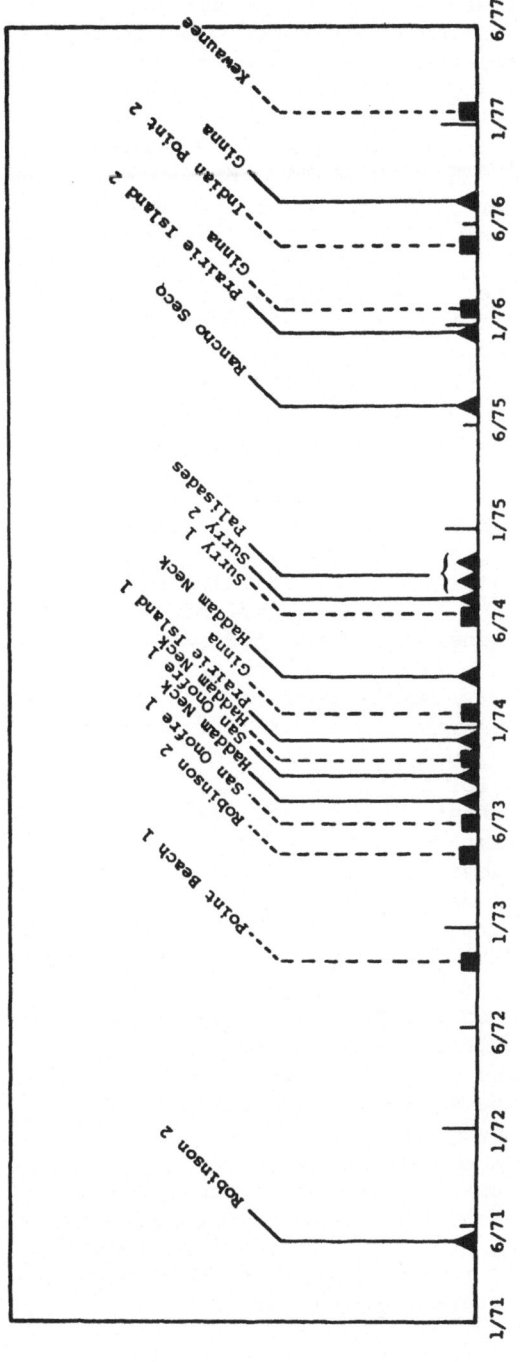

Figure 16. Major Turbine Blade Failures Causing Extended Nuclear Plant Outages (Other turbine or generator outages are not included.)

b. Tube vibration caused by cross flow
    from recirculating water has caused
    tube fretting at the support plate
    region and bend region.

c. Corrosion in the tube sheet crevice
    region, the classical concern of
    designers, has caused only a limited
    number of failures.

d. Condenser tube failures can be the
    cause of the introduction of im-
    purities into the feedwater, thus
    perturbing the sensitive balance of
    steam generator water chemistry and
    leading to accelerated corrosion.

e. Secondary water chemistry control
    has been a point of major discussion
    in the prevention of steam generator
    tube failures. In 1974, the high
    incidence of wastage corrosion in
    steam generators prompted most sup-
    pliers to recommend a switch in
    secondary chemistry from phosphate
    treatment to all volatile treatment
    (AVT). Phosphate treatment has been
    blamed for heavy sludge formations
    in stagnant flow areas, leading to
    tube wastage or thinning. However,
    a weakness of the AVT method is that
    it does not neutralize the attack of
    contaminants from condenser tube leaks.
    In addition, Surry 1 and Surry 2 (see
    Section IV) and Turkey Points 3 and
    4, all of which have been switched to
    AVT, have experienced significant
    swelling of carbon steel tube sheets
    which have caused "denting" of tubes
    and subsequent tube leaks.

    B & W has a different steam generator
    design and has always specified a high
    purity, all-volatile chemical treat-
    ment of the feedwater of their once-
    through-steam-generator (OTSG). In

addition, B & W recommends high-
purity feedwater from 100% full-flow
condensate polishing demineralizers.
However, the B & W Oconee steam generators have begun to show signs of
an increasing number of tube defects
that resulted in a number of outages
for tube plugging.

Figure 17 summarizes the plants that have encountered failures of steam generator tubing in each year. In
many of the plants (e.g., Surry 1, Surry 2, Turkey Point 3,
Turkey Point 4) there have been multiple outages within a
single year due to steam generator repairs; however, they
are represented in this simple display as the total number
of failures in a given year. Based on the individual plant
operating experience data, one can note that during 1974,
most plants had switched from phosphate treatment of the
secondary water to an all volatile chemistry (e.g., hydra-
zine). From Figure 17, this switch coincided with the re-
porting of fewer wastage failures and the emergence of the
problem referred to as "denting".

Also from Figure 17, there is an apparent high
frequency of Westinghouse steam generator tube problems.
This can be explained in part by the much larger population
of installed Westinghouse steam generators. Figure 18 shows
a composite plot of the contribution of steam generators to
major outages, broken down by principal suppliers. Westinghouse units have accumulated the largest amount of operational experience and the highest amount of outage time attributed to steam generators. B & W and CE units have a much
smaller amount of operational experience, since there are
only six B & W units and six CE units included in these
data. Therefore, if we use major outage time on a per plant-
year basis as a measure of performance, the Westinghouse
steam generators are performing as well or better than the
B & W and CE units.

The distribution of major outages with plant age
is similar for Westinghouse and B & W units, despite their
apparent marked difference in design. Each has experienced
a dramatic increase in outage time in the fourth year of
commercial operation. Less confidence can be applied to the
CE distribution, since all major outage time is associated
with a single plant, Palisades.

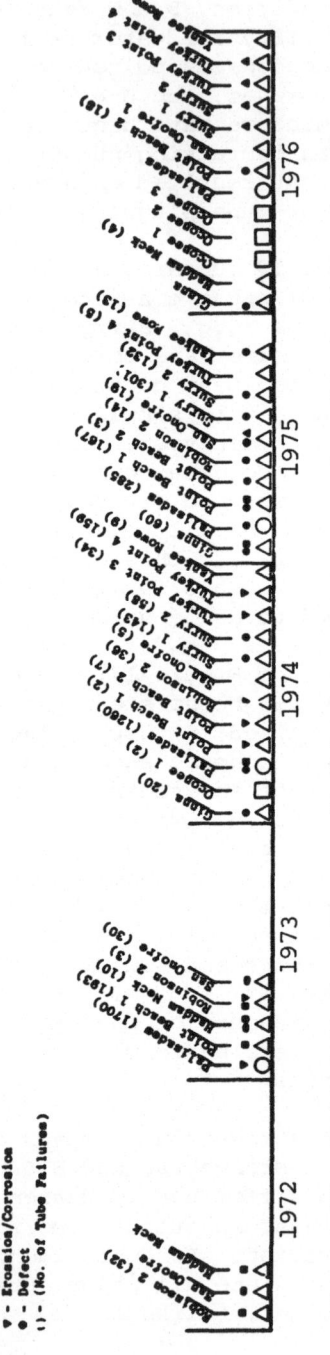

Figure 17.   United States PWRs Requiring Steam Generator Tube Plugging by Year of Event
(Vendor, Reported Failure Mode, and Number of Tubes Repaired Included)
(See References 27, 28, 29, 30, 31)

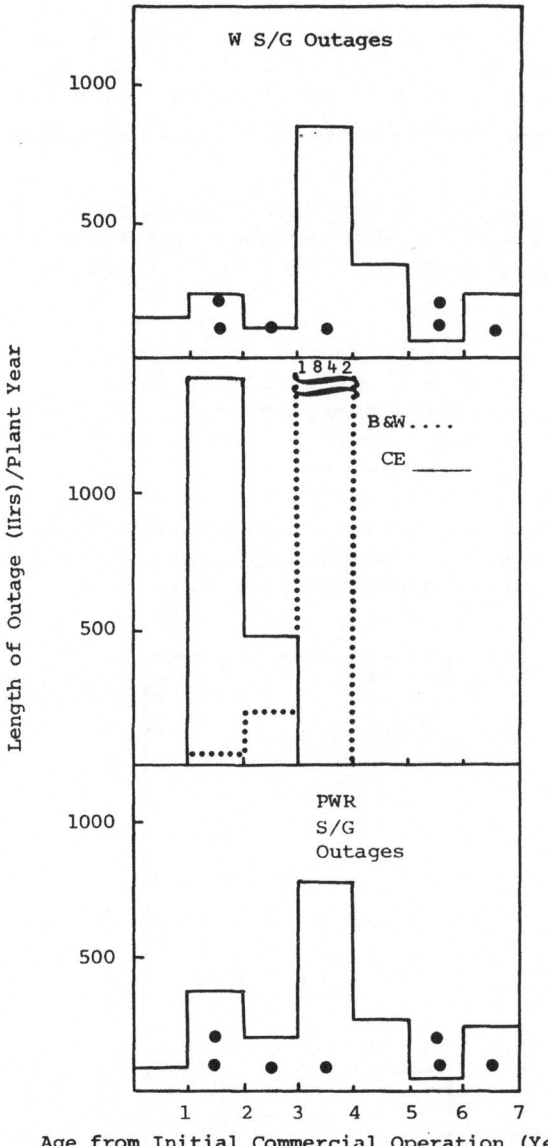

Figure 18.   Comparison by Vendor of Major Outage Variations
             Related to Steam Generators

A very limited data sample beyond the fourth year
(virtually no B & W or CE experience) results in more uncertainty in the fifth, sixth and seventh year average outage
numbers; however, the outages due to steam generators in
those years represent a significant contribution to the
overall outage rate and are a principal area of concern in
long-term PWR plant performance.

3.   Outage Trends in Pumps.   Reactor coolant pumps in
PWRs and recirculation pumps in BWRs have been the cause of
a number of major outages, particularly early in plant life
(see Figure 19).

4.   Outage Trends in Cores.   Reactor core-related
problems have tended to be of a generic nature; generally,
they are caused by design errors incorporated into several
units.   Most core-related repairs were performed during refueling operations.   For those cores between refuelings,
extended outages were required to modify the design.   The
actual failure mechanism has been flow-induced vibration.
Note that these problems characteristically have surfaced
during the first two years of operation (see Figure 20).

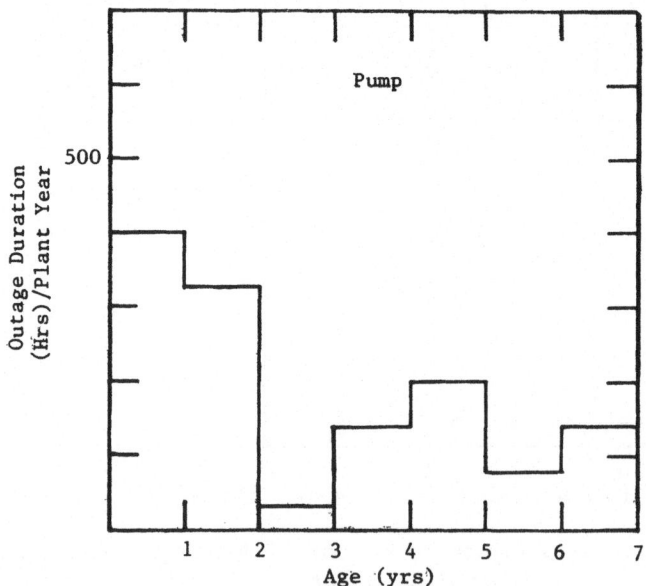

Figure 19.   Trend in Outages for Pump-related Outages

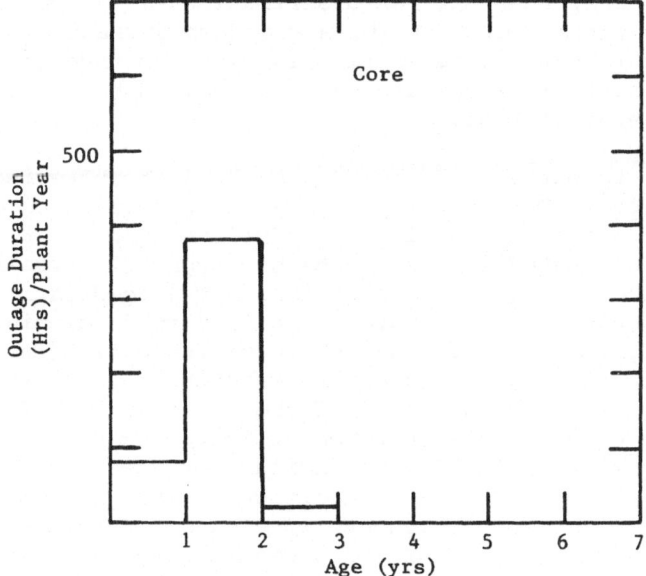

Figure 20.   Trend in Major Outages for Core-related Problems

C.   Comparison of PWR and BWR Refueling Outage Trends
      Versus Plant Age

      As noted in Section II B, there is a wide variation in
the details of plant design among the operating reactors.
One general division that can be interpreted easily and with
little ambiguity is the distinction between PWRs and BWRs.
As is evident from the previous section, certain aspects of
PWR and BWR design are reflected directly in the types of
outages and the time requirements for outages in PWRs and
BWRs.   Therefore, this subsection presents a comparison of
the two reactor types in order to distinguish any fundamental
differences in their respective operating histories.

      A comparison of the mean refueling outage lengths for
BWRs and PWRs indicates an 18% difference in the average refueling outage length for the two distributions.   The distributions of all BWR refueling outages has a higher mean

value (12.3 weeks) than the PWR distribution (10.4 weeks)
and a broader variation; that is, a standard deviation of six
versus five for PWRs. This section is aimed at understanding the reason for this difference in refueling outage length
for the two types of LWRs.

In order to better understand the trend of refueling
outage duration in PWRs and BWRs, consider a comparison of
refueling outages as a function of refueling number. Since
we are most interested in the trends of "typical" refueling
outages and there have been some infrequent long-duration
outages, the method used in this comparison is to eliminate
the highest duration outage in each refueling cycle (if it
exceeds four months) as being non-representative of refueling
outages. Figure 21 shows a plot of the distribution of average refueling outages as a function of refueling number. The
length of the first refueling is similar for PWRs and BWRs
(within 5%), and as discussed in Section III A, it is significantly longer than subsequent refuelings. For both
PWRs and BWRs, the lengths of the second through the fifth
refuelings* are nearly constant, indicating that after the
first refueling, the plant can be considered as having
reached refueling maturity and to have achieved an approximately constant outage time per refueling as is shown in
Figure 21. The BWR plateau is 14% to 20% higher than an
average PWR plateau, indicating that mature BWR plants may
require more time to allow completion of all the necessary
work items scheduled during refueling outages than in PWR
plants. In Section III A, it is shown that the actual time
spent for required refueling operations is comparable for
BWR and PWR plants; therefore, the differential is in the
required plant maintenance. This can be explained in part
by the larger number of generic problems that BWR plants
have encountered thus far, which have forced extensions of
the refueling outages (e.g., feedwater sparger and core spray
problems, core internal modifications, replacement of mar-
ginal performance fuel that requires critical path time to
identify).

A comparison of refueling outages for BWRs and PWRs

---

*   The sixth and seventh refuelings have a very small
     population, and consequently, the statistics are
     inadequate to say anything meaningful about the trends.

from Figure 21 results in the following conclusions (see also, Table VI):

1. The average BWR refueling outage is longer than the average PWR refueling outage.

2. BWR and PWR first refuelings have approximately the same average length.

3. It appears that the refueling length is approximately constant after the first refueling of a plant, and that the average PWR refueling outage may be less than the average BWR refueling outage by approximately 15%.

4. The average refueling length for a mature PWR plant is approximately 1 1/2 weeks less than the comparable BWR outage length, as shown below:

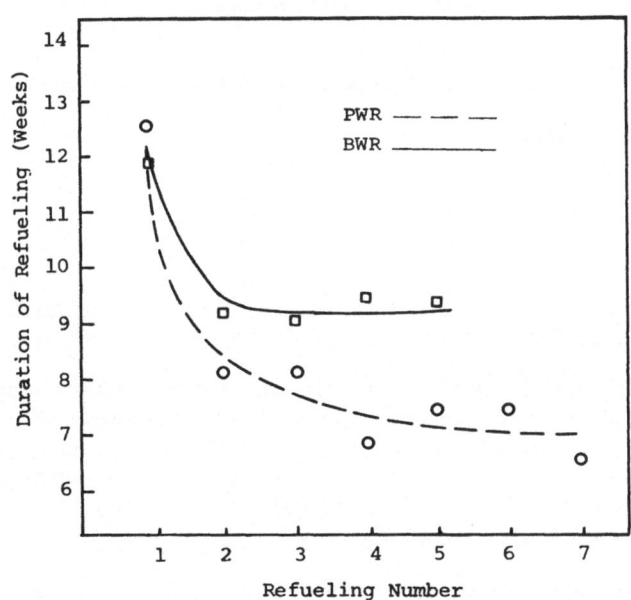

Figure 21.   Refueling Duration as a Function of the Number of Refuelings

TABLE VI

COMPARISON OF AVERAGE REFUELING LENGTHS
Including All Refuelings Except the Initial

	Average Refueling Length of Mature Plant (Weeks)
All Plants	8.8[†]
BWRs[*]	9.2[*]
PWRs[*]	7.8[*]

[*]The longest refueling outage, if greater than 4 months, has been eliminated in the assessment of each refueling cycle average

[†]All refuelings greater than 3 months have been eliminated

D. Variation of Refueling Outage Length as a Function of Plant Size

One aspect of utility planning for new plant capacity is the determination of the optimal plant size. The selection of plant size is based upon an assessment of many variables; including the amount of capital investment, the potential economy of scale, the projected energy needs, and the expected plant availability.

Past studies have attempted to evaluate the variation of plant availability as a function of plant size; however, the limited data and the large variability in the data have made the correlations statistically meaningless. This sec-

tion takes a look at one contributor to plant availability-
the refueling outage. Since refueling outages are major
contributors to plant unavailability, this section seeks to
determine if there is a positive correlation between refueling outage length and plant size. The determination of the
variation in refueling outage duration versus plant size
(rating) can be useful input to improved planning. The
analysis performed here is based upon data from the initial
seven years of plant operation (less than one-fifth of the
projected plant life). As previously mentioned, these data
are highly biased, since 35% of the data points are first
refuelings. Given these two strong limitations, a statistical analysis of the data was performed which indicates
with 95% confidence that there is a small, positive correlation between plant size and the length of the refueling
outages. One test of the validity of this conclusion is to
look at the lower bounds on the refueling outage length as
a function of plant size. It can be seen that the minimum
time to refuel plants shows a trend similar to that for the
best estimate slope. In other words, the minimum refueling
outage time for both PWRs and BWRs as a function of plant
size is a monotonically increasing function.

A similar conclusion also can be gleaned from looking
at the longest refueling outage lengths. However, the
longest refuelings show much more variability, indicative
of the fact that the longer refueling outages are, in general, dominated by a major maintenance item such as turbine
repair, plant modification, in-core repairs, etc.

A linear regression analysis of the data for PWR and
BWR plants indicates that the best fit approximation to the
data has a positive slope. In order to show that this correlation holds with a high level of confidence (95%), consider the sample regression calculation in Table VII for
BWR plants.

Figure 22 illustrates the 95% confidence interval on
both:

1.   A single future value of Y

2.   On the calculated regression line

Note that the confidence interval on the regression line is

much tighter than the confidence interval on a single future
prediction. It is the confidence interval on the regression
line on which we will concentrate for the subsequent discussion.

These calculations show that the slope of the trend
line for variation in refueling outage length versus plant
size is positive with greater than 95% confidence. As an
example, one can calculate the outage length differential
between a refueling occurring in the average 500 MWe plant
versus that in the average 1,000 MWe plant. Table VIII compares the refueling outage length differential for the best
estimate case and the 95% confidence interval on the slope
of the regression line.

TABLE VII

BWR Regression Correlation Between Plant Size
and Refueling Outage Length

$\hat{\beta}$ = estimate of the slope of the best fit = 1.19×10^{-2} weeks/MWe

σ_y = standard deviation of the refuelling duration = 5.16 weeks

σ_x = standard deviation of the plant size = 155 MWe

N = number of data points (plants = 46

σ_β = standard deviation of the estimate of the slope = $\sigma_y / \sigma_x \sqrt{N}$

   = $5.16 / 155 \sqrt{46} = 4.91 \times 10^{-3}$ weeks/MWe

γ = 95% confidence limit on the slope = $\hat{\beta} \pm 1.65 \, \sigma_\beta$

   = $1.19 \times 10^{-2} \pm 1.65 \times 4.91 \times 10^{-3}$

   = $(3.8 \times 10^{-3}, 1.99 \times 10^{-2})$ weeks/MWe

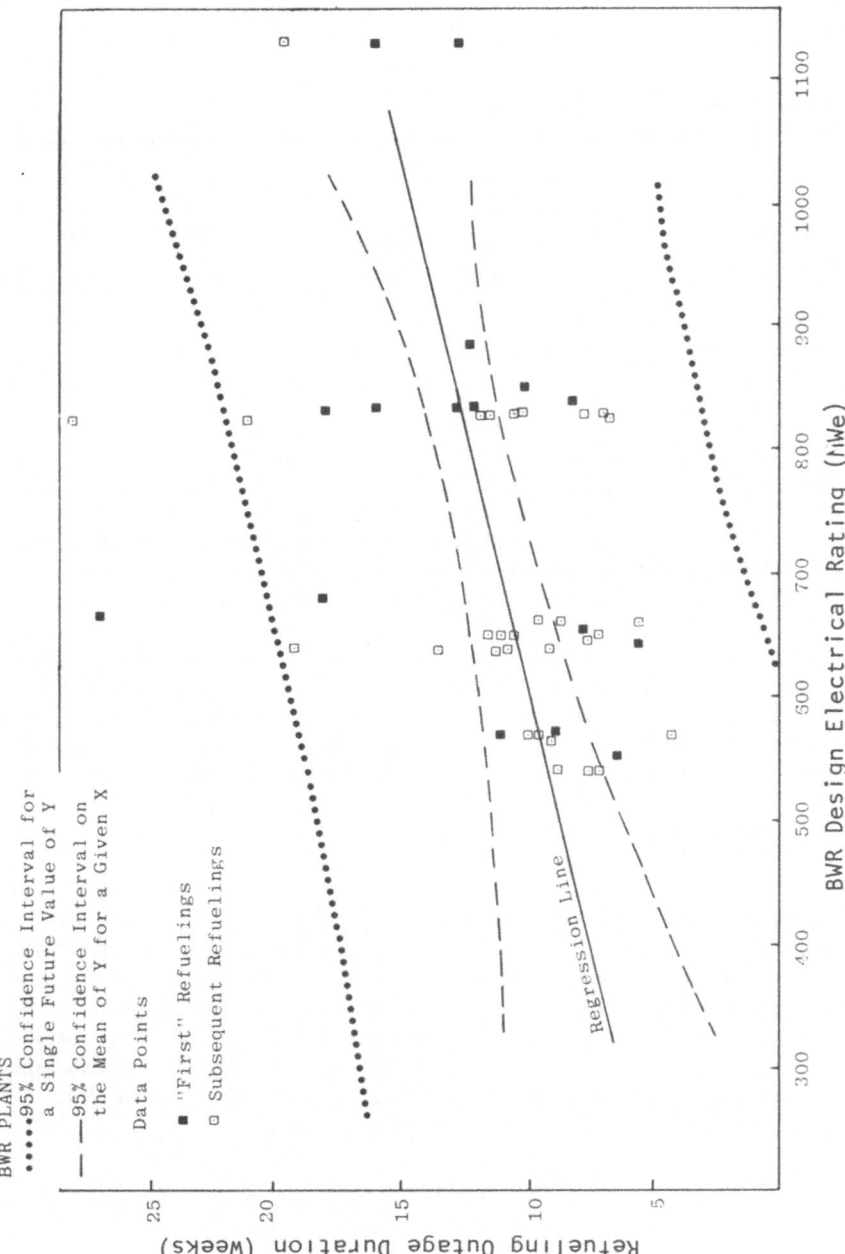

Figure 22.   Variation of Refueling Outage Duration as a Function of Plant Rating for BWRs

TABLE VIII

Refueling Outage Projections for "Typical" BWR Plants of
Two Sizes

BWR Plants	Confidence Level	Refueling Outage 500 MWe Plant	Refueling Outage 1000 MWe Plant
High Estimate	95%	Base	Base + 10 weeks
Best Estimate		Base	Base + 6 weeks
Low Estimate	95%	Base	Base + 2 weeks

The best estimate regression indicates that a 1,000 MWe BWR plant may take six weeks longer to refuel than a 500 MWe BWR. There is a great deal of scatter in the data, which results in a large uncertainty for this regression model to predict precise numbers; however, as shown in the table, the operating experience data indicate a trend that large BWR plants take longer to refuel than smaller units.

The same calculation can be carried out for PWR plants as shown in Table IX (see Figure 23). Again, the slope of the correlation is positive within 95% confidence limits; however, the absolute value of the slope is less than that determined for BWR plants. The PWR outage differential between 500 and 1,000 MWe plants predicted for this correlation is summarized in Table X.

It must be noted that the above regression analysis for PWR plants yields a minimum slope that is zero for all intents and purposes. This minimum slope would indicate that there is no correlation between PWR size and refueling outage length. However, the above comparisons, as illustrated in Figures 22 and 23, indicate that there is a high probability that the calculated linear regression of the trend of refueling outage duration as a function of plant size has a positive slope for both BWR and PWR plants.

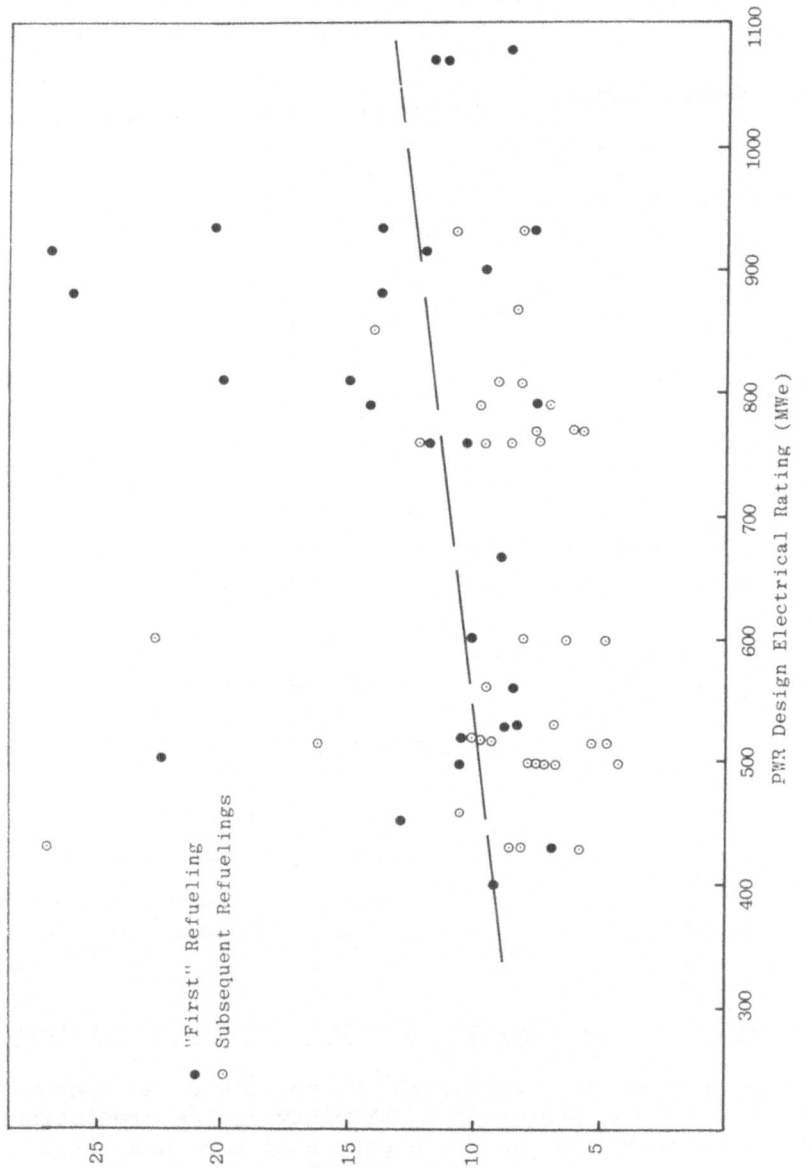

Figure 23. Variation of Refueling Outage Duration as a Function of Plant Rating For PWRs

TABLE IX

PWR REGRESSION CORRELATION
Between Plant Size and Refueling Outage Length

$\hat{\beta} = 6.25 \times 10^{-3}$ weeks/MWe

$\sigma_y = 5.29$ weeks

$\sigma_x = 176$ MWe

$N = 65$

$\gamma = 6.25 \times 10^{-3} \pm 1.65 \ (5.29 \ / \ 176\sqrt{65})$

$= (9.4 \times 10^{-5}, \ 1.24 \times 10^{-2})$ weeks/MWe

TABLE X

REFUELING OUTAGE PROJECTIONS
For "Typical" PWR Plants of Two Sizes

PWR Plants	Confidence Level	Refueling Outage Duration 500 MWe Plant	Refueling Outage Duration 1000 MWe Plant
High Estimate	95%	Base	Base + 6.2 weeks
Best Estimate		Base	Base + 3.1 weeks
Low Estimate	95%	Base	Base + .1 weeks

E.   Accuracy in Predicting the Length of a Refueling Outage

The purpose of this subsection is to provide an approximate measure of the degree of uncertainty in the prediction
of refueling outage lengths. The data used for this calculation are from the utility's estimate for the projected
length of the refueling that is given to the NRC[20], versus

the actual length. Only data on approximately 60% of the
total refuelings considered in this report are available
for use in this section; therefore, the statistics are drawn
from a smaller, overall population. In addition, the reader
is cautioned to note that the estimate of outage duration can
vary during the planning stages as new maintenance items,
inspections, or tests are identified. For consistency, the
projected refueling length used in this section is that provided by the utility to the NRC just before the outage began.

Figures 24 and 25 provide a simple comparison of the
estimated refueling outage duration versus the actual length
of PWR and BWR refueling outages. The conclusions from this
comparison are the following:

1.   Predictions of the longest refueling outages
      (i.e., those that contain a significant
      amount of critical path maintenance) tend
      to have a much larger uncertainty than "pure"
      refuelings. In virtually all cases, the
      uncertainty is in the direction of under-
      estimating the length of the outage.

2.   First refuelings are the longest, and have
      the highest deviation from the estimated
      length.

In order to further characterize the accuracy of the
predicted refueling outage length, Figures 26 and 27 give a
frequency histogram distribution of the differential between
the actual refueling time and that estimated by the utility.
The differential between the actual and estimated refueling
outage time can be: (1) positive if the length of the refueling outage has been underestimated; (2) negative if the
length of the refueling outage has been overestimated.

For BWR plants, data are available for 28 refuelings[20]
that compare the utility estimate of the refueling outage
with the actual time required. In general, there is a wide
variability in the utility estimates versus the actual outage length (see Figure 26). The following are additional
conclusions for BWRs:

1.   Major maintenance/repair discovered during
      refueling can significantly extend the outage

far beyond that anticipated (e.g.,
feedwater sparger repairs, recirculation by-pass line repair).

2.   Even when maintenance is anticipated,
the complexity of the required tasks
tends to be underestimated.

3.   In general, the outages are underestimated consistently by approximately two to three weeks.

For PWRs, data are available for 50 refuelings(20),
which compare the projected versus the actual refueling outage durations (see Figure 27). The conclusions from this
comparison for PWRs is similar to those cited above for
BWRs, with the following exceptions:

1.   The median refueling underestimate for the
PWRs is between one and two weeks.

2.   The PWR distribution has approximately the
same number of large underestimates of refueling outage length as in the BWR distribution; however, the percentage effect
is much smaller because of the larger amount
of PWR refueling outage data points.

3.   The PWR distribution has a smaller standard
deviation than the BWR distribution, indicating a slightly greater consistency in
the estimated versus actual refueling outage
length.

All of the above effects are subject to wide variability
that can be caused by unexpected equipment failures, differences in planning and procedures, the quality of the maintenance personnel, and many other items. Therefore, conclusions
based upon the accuracy of refueling outage predictions are
of use in terms of informing utilities, vendors, and AEs of
the necessity of: (1) planning outages carefully; (2) antic-
ipating major maintenance efforts based upon operating experience at other plants; and (3) allowing for adjustments to
meet load requirements in the event of extended outages.

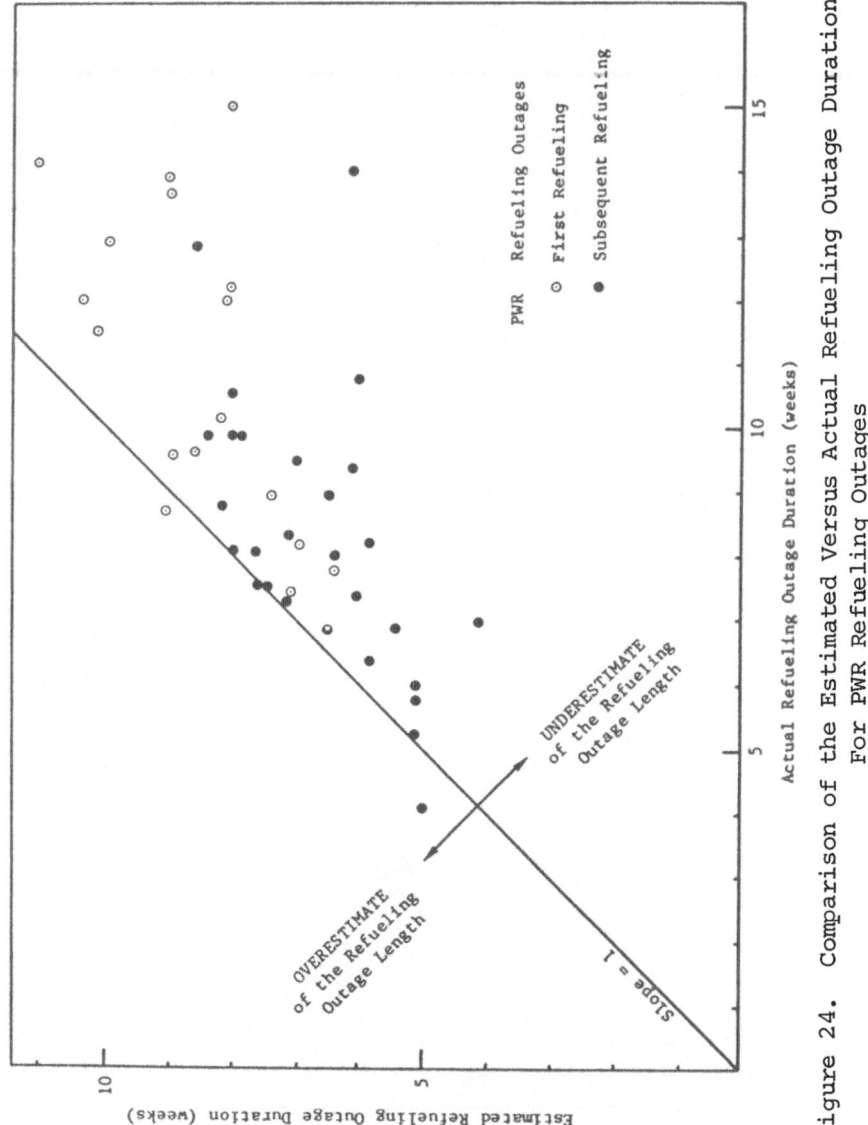

Figure 24. Comparison of the Estimated Versus Actual Refueling Outage Duration
For PWR Refueling Outages

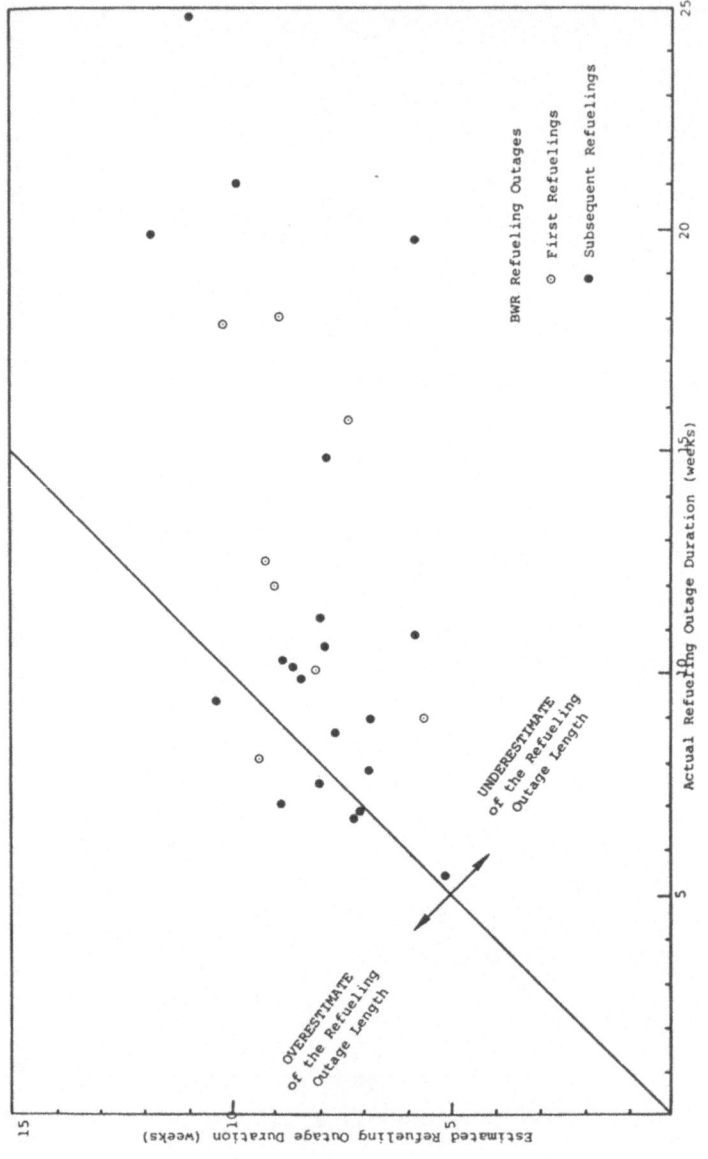

Figure 25.   Comparison of the Estimated Versus Actual Refueling Outage Duration
For BWR Refueling Outages

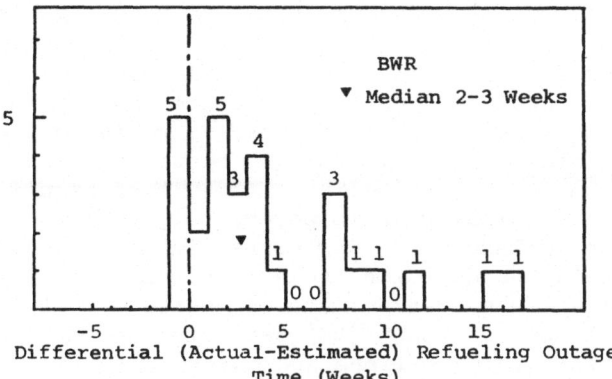

Figure 26.   Accuracy of BWR Refueling Outage Predictions:
A Frequency Histogram of the Difference Between
the Actual and Estimated Outage Time

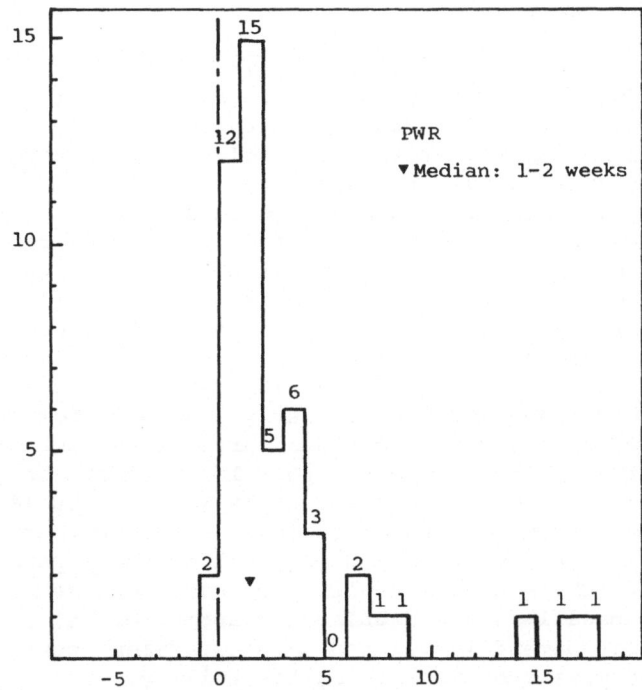

Figure 27.   Accuracy of PWR Refueling Outage Predictions:
A Frequency Histogram of the Difference
Between Actual and Estimated Outage Time

IV.   INCREASING THE LENGTH OF THE REFUELING CYCLE

The present annual refueling cycle has evolved as a trade-off between fuel enrichment costs and the frequency of refuelings, using an anticipated refueling outage length of one to three weeks. Based upon operating experience, which indicates that refueling outages take significantly longer than originally anticipated, coupled with the rapid increase in replacement power costs, it appears desirable to reevaluate the decision to have approximately annual refuelings. The operating data demonstrate that refueling outages include more than just refueling operations, they include testing, inspections and maintenance items. The result is that refueling outages have taken anywhere from four weeks to more than 20 weeks, depending upon the work items required. Previous estimates of refueling outages were the result of over-optimism on the speed with which a refueling could be performed, and the fact that all the operations occurring during a refueling outage were not in-cluded in the previous estimates.

A constraint on the overall length of ideal refueling cycles is that they should be integral units of six months. Since the nation as a whole has its peak energy require-ment during the summer (followed closely by winter elec-trical requirements), the greatest percentage of electric generating capacity is planned to be available during the summer and winter months. For the summer peak periods, the capacity planned to be unavailable due to scheduled maintenance, averages less than 2% (summers of 1976 to 1979)(21); the different electric reliability councils re-porting a range from essentially zero to a maximum of only 4%. The United States average for the winter peak periods is about 9%; however, particularly severe winter weather in the midwestern and northeastern parts of the country can shrink the reserve to near zero. In any case, both summer and winter demand periods are times when the nuclear plant power generation is most necessary; therefore, utilities are anxious to avoid any outages during these peak power periods. Therefore, refueling outages, because of their length, usually are scheduled for spring or fall months when power demand is reduced and replacement power can be provided more economically by the utility.

As mentioned in the foregoing discussion, a potential method of improving plant availability is an extension of the refueling cycle from the present 12 months to 18 months. The chief advantage of this concept is that it can provide an increase in the plant availability by reducing the scheduled outage time; this translates directly into increased plant productivity. Specifically, there is an increase in plant availability. This increase in availability of 6% per year represents a potential saving of five to ten million dollars per plant in replacement energy costs per year. Therefore, the potential savings in lengthening the refueling cycle is significant.

In the idealized 18 month cycle, the major assumption is that there is no increase in the fraction of outage time occurring outside refueling outages. This is shown to be good assumption since at present for a mature plant, there are approximately 500 hours per plant year outage time occurring outside refueling outages. This translates into 750 hours between refuelings which occur on an 18 month refueling cycle. Operating experience data for plants that have 18 month refueling intervals seem to confirm that this is sufficient time to cover maintenance items that cannot be properly attended to during the less-frequent refuelings. Since most utility maintenance programs are geared to the annual refueling cycle, additional effort may be required to attempt to optimize a preventive maintenance program around the longer 18 month cycle. This optimization may lead to the conclusion that some preventive maintenance still should be performed on an annual schedule during other forced outages or during scheduled shorter duration outages between refuelings.

A key assumption in the extension of the refueling interval is that the fuel performance is adequate to allow operation over 18 month periods without forcing plant deratings to limit off-gas activity or primary coolant activity. Current information(22) suggests that the newest fuel design for both PWR and BWR plants have substantially reduced fuel leakage rates; therefore, presently available fuel performance appears adequate to meet the longer cycles.

Because extending the refueling cycle from 12 months to 18 months requires possible changes in fuel enrichment, the decision to switch to the longer cycle must be accom-

panied by long-range planning as to the fuel loadings that
will be adopted and the adequacy of enrichment capability
needed to meet the demand. Typical lead times in specifying
fuel enrichment are on the order of two to three years prior
to fuel loading.

In addition to the empirical evidence from the operating data for potential increased availability using longer fuel cycles, there are some utilities that have examined
the problem, judged the 18-month refueling cycle to be superior to the annual refuelings, and already have made the
switch. Surry 1 and Surry 2 now are on the first of their
18 month cycles; Nine Mile Point, Quad Cities, Brown's Ferry
1 and Brown's Ferry 2 are among those that have implemented
on an 18 month cycle. Other units where longer cycles have
evolved are Connecticut Yankee and San Onofre, which are
Westinghouse reactors of similar design using stainless steel
clad fuel rods; therefore, these units are slightly atypical
of other reactor plants that use zircaloy fuel cladding.

As with any decision affecting a nuclear power plant,
changes in the length of the refueling cycle affect a wide
variety of considerations. Since many nuclear facilities
face unique problems or obstacles, the following discussion
of various aspects of the impact of changing the fuel cycle
on nuclear power plant operation is included to emphasize
the many interrelated factors that must be considered before
a decision is made to extend the refueling cycle.

From the standpoint of advantages of increasing the
interval length between refuelings, we can cite the following potential advantages:

1.   Increased plant availability. The elimin-
      ation of one refueling outage every three
      years translates into an increase of 6% in
      plant availability. A 6% increase in plant
      availability translates into savings of
      replacement power costs of approximately
      five to ten million dollars per year!

2.   Greater flexibility in fuel cycle. Planning
      for 18 month fuel cycles will enable a plant
      that is running well to stay on-line and
      available to produce base load power. Only

"bad" things can happen when a plant is
shut down. This is a subjective judge-
ment, based upon a review of operating
experience which indicates that it is
desirable to maintain a smooth-running
plant at power as long as possible. The
longer cycle will not preclude early
shutdown for removal of possibly leaking
fuel or other major maintenance. In other
words, greater system flexibility is
afforded by the change to longer cycles.

3. Reduction in the number of heat-up,
    cool-down transients. An important
    variable in determining a nuclear power
    plant's lifetime is the number of large
    temperature transients from normal
    operating temperature to "cold iron"
    and back. A reduction in the number of
    refueling operations should result in
    a decrease in the number of such tem-
    perature cycles. This would be a
    direct benefit in terms of component
    lifetime.

4. Plant security. This is an important
    aspect of nuclear power generation from
    the standpoint of public health and
    safety and the protection of expensive
    equipment. A decrease in the frequency
    of refueling outages will result in a
    reduction of one-third in time of ex-
    posure of the plant to outside contrac-
    ting personnel and even in-house main-
    tenance personnel in vital areas of the
    plant, particularly inside containment
    (e.g., reactor vessel, control rod drive
    mechanism, main coolant piping). In
    addition, as concern over safety, safe-
    guards, and prevention of sabotage in-
    creases, the restrictions on maintenance
    personnel will increase even more, re-
    sulting in an increase in the already
    staggering administrative constraints
    on workers and their actions, and there-

fore on the length of refueling outages.
A decrease in the number of refuelings
will lessen the burden on the utilities
in dealing with the administrative work-
load involved in ensuring that safeguard
requirements are met during such an extended maintenance.

5.   Reduction in the amount of manpower and
     plant time for regulatory review. The
     18 month refueling cycle has another
     advantage: that of reducing the number
     of times the utility must interface with
     the NRC. Historically, each regulatory
     review has demonstrated the potential
     for creating additional back-fit require-
     ments. Because startup from a refueling
     receives increased NRC attention, each
     recovery from refueling requires a sig-
     nificant effort on the part of plant
     management to satisfy NRC that the plant
     is being operated and maintained safely.
     If a utility can minimize these inter-
     faces with NRC, plant management and
     operational personnel will have increased
     time available for plant operational needs.

6.   Reduction in the plant personnel radiation
     exposure. An additional argument for the
     increase in refueling intervals comes not
     from an availability argument but from a
     consideration of the number of men exposed
     to the increased levels of radiation involved
     in refueling operations. The current NRC
     regulations require a compliance with "as
     low as reasonably achievable" philosophy.
     Historically, most radiation exposure at
     nuclear power plants occurs during refuel-
     ing and maintenance operations. A reduction
     in the number of refuelings by increasing
     the interval between them could mean a re-
     duction in the total man-rem exposure of
     operations and contractor personnel.

There are, however, a number of variables that may detract from the advantages of increasing the refueling cycle length. Two examples of areas that require an evaluation by each utility with regard to its own particular needs are the following:

Increased enrichment versus larger batch size. The principal trade-off in deciding whether to extend the refueling cycle length is in the method of ensuring adequate fuel to extend the cycle an additional six months. Two possible alternatives are:

1. increasing the enrichment of the fuel assemblies

2. increasing the batch size exchanged during each refueling

The first alternative, increasing fuel enrichment, has several uncertainties involved in it that require additional information. In particular, higher fuel enrichment may lead to: (1) higher fuel burnup, which has not been shown to be consistently achievable to date; (2) higher enrichment cost, associated with the required higher enrichment; (3) higher power peaking, which could reduce the core thermal margin. The second alternative, increasing the batch size, would avoid significantly higher burnups, but would entail a more involved fuel management scheme to optimize performance. The batch size probably would be increased from one-third to two-fifths of a core. For both alternatives, the fuel performance over an 18 month calendar period is of crucial importance.

A side issue that is not adequately discussed in the literature is the question of reduction in the number of spent fuel assemblies that must be handled in the back end of the fuel cycle. If the choice is made to increase the initial enrichment there will be a decrease in the number of fuel assemblies required, and therefore, the number of spent fuel assemblies that eventually must be disposed of over the life of the plant. This change is independent of the choice to lengthen the refueling cycle; however, it could be accomplished simultaneously.

<u>Increased forced outage time.</u> The unknown in this evaluation is whether increasing the length between refuelings will lead to an increase in the forced outage time to perform needed maintenance or repair. Because experience is lacking in this area, it must be considered an unknown factor at this time.

V. CONCLUSIONS

Previous studies of nuclear plant outages have emphasized the necessity of planning in order to minimize outage time. The facts in this report are presented to aid in that planning by increasing the understanding of key decision makers about the makeup of major nuclear plant outages. Because each plant or small group of plants may have unique problems, it is difficult to formulate generalized solutions for the entire industry by looking at only one or two plants. Therefore, this report summarizes a general review of all United States nuclear plants in an attempt to establish a pattern in the industry. The following conclusions are reached from the assessment of this population:

1. Refueling outages account for approximately 60% to 75% of the unavailability of a mature nuclear plant.

2. Of the total refueling outage time, refueling operations required to insert new fuel in the reactor occupied only 50% of the time. The remaining time is required for test, inspection and maintenance.

3. Refueling outages are, in general, underestimated by one to three weeks.

4. The impact of refueling outages on plant performance may be reduced if the refueling cycle is increased from 12 to 18 months.

5. There are certain selected classes of plant equipment that have been the cause of a significant proportion of all major outage time. In particular, the highest contributor to lost productivity time is associated with turbine blading problems, steam generator tube leaks and pumps.

Since each utility has unique needs, it is important for the utility to evaluate the advantages of switching to a longer refueling cycle versus the potential drawbacks for its own particular situation; i.e., generator mix, replacement power costs, maintenance crew support, fuel management scheme, historical plant performance, fuel performance.

The extensions of this effort are in the precise method in which it is applied to a utility. In particular, the preventive maintenance schedule is of crucial importance in maintaining or reducing the current level of outage time outside refuelings. The preventive maintenance, however, must be planned to be compatible with the revised refueling schedule, and with equipment and technical specifications.

REFERENCES

1. Komanoff, Charles, Power Plant Operating Reliability, Council of Economic Priorities, July, 1976.

2. Bock, J. T., An Analysis of Factors Affecting Capacity Factors of Nuclear Power Plants, California State Energy Conservation and Development Commission, August, 1976.

3. Andrews, H. N., et al, "Optimizing the Refuel Cycle: Two Views," Nuclear News, pp 71-76, September, 1973.

4. BWR Power Plant Training, NEDO-10260, October 1971.

5. Long, R. L., et al, "Implementation of Availability Engineering in the Nuclear Utility Industry," Transactions of the American Nuclear Society, Vol. 27, Page 96, November, 1977.

6. Bernath, L. and Latham, D. W., "Sundesert Reliability Engineering," Transactions of the American Nuclear Society, Vol. 27, Page 97, November, 1977.

7. "A Summary of Nuclear Power Plant Operating Experience for 1975," Electric Power Research Institute, EPRI NP-263, October, 1976.

8. Komanoff, Charles, "Power Plant Operating Reliability" (Draft), Council on Economic Priorities, July, 1976.

9. Lapides, M., and Zebroski, E., "Use of Nuclear Operating Experience to Guide Productivity Improvement Programs," Electric Power Research Institute, SR-26R, November, 1975.

10. Bridenbaugh, D. G. and Burddsall, G. D., "Application of Plant Outage Experience to Improve Plant Performance," Proceedings of the American Power Conference 1974, Vol. 36, pp 143-150.

11. Lapides, M. E., "Better Operating Data Help Improve Productivity," Nuclear Engineering International, pp 53-55, October, 1976.

12. Edwards, D. W., "Outage Management: A Case Study," Nuclear News, pp 45-49, December, 1975.

13. Wandke, N. E., "Maintenance Outage Management at Commonwealth Edison," Nuclear Engineering International, pp 55-57, April, 1977.

14. Haig, R. and Kuffer, K., "Application of a Feedback Control System in Switzerland," Nuclear Engineering International, pp 57-60, April, 1977.

15. "Steam Generators to be Replaced at Surry and Turkey Point Nuclear Units," Nucleonics Week, No. 22, Vol. 18, June 2, 1977.

16. Stevens-Guille, P. P., "Steam Generator Tube Failures: World Experience in Water-Cooled Nuclear Power Reactors During 1972," Nuclear Safety, No. 3, Vol. 16, May-June, 1975.

17. Stevens-Guille, P. P., and Hare, M. G., "Steam Generator Tube Failures: World Experience in Water-Cooled Nuclear Power Reactors in 1973," Chalk River Nuclear Laboratories, AECL-5013, January, 1975.

18. Hare, M. G., "Steam Generator Tube Failures: World Experience in Water-Cooled Nuclear Power Reactors in 1974," Nuclear Safety, No. 2, Vol. 17, March-April, 1976.

19. Hare, M. G., "Steam Generator Tube Failures: World Experience in Water-Cooled Nuclear Power Reactors in 1975," Nuclear Safety, No. 3, Vol. 18, May-June, 1977.

20. Operating Units Status Reports, Nuclear Regulatory Commission, NUREG 0200.

21. "Electrical Generating Plant Availability," Staff Report, Federal Power Commission, Bureau of Power, May, 1975.

22. Roberts, J. T. A., et al, "Planning Support Document for the EPRI Light Water Reactor Fuel Performance Program," Electric Power Research Institute, EPRI-NP-370-SR, January, 1977.

SYNERGETIC NUCLEAR ENERGY SYSTEMS CONCEPTS

A. A. Harms
McMaster University, Hamilton, Canada

I.   INTRODUCTION

The early developmental period of the civilian nuclear power program was generally marked by considerable optimism about this new-found energy source. This optimism not withstanding, it nevertheless had been recognized that, in the long term, reactors based on thermal fission would need to be augmented - or possibly even supplanted - by other nuclear energy technologies. This view was imposed by the realization that the terrestrial abundance of fissile materials was insufficient to support an expansive thermal fission economy.

One consequence imposed by the acknowledged scarcity of fissile materials was an emerging emphasis on two nuclear technologies which promised to eliminate the fuel exhaustion problem; the fast-breeder reactor and the fusion reactor were identified as concepts worthy of major research and developmental emphasis. Once started with an initial inventory of fissile material, the fast-breeder would eventually require only fertile fuels which it would transmute into fissile materials not only for itself but also for other thermal fission reactors while the fusion reactor, once fully developed, would burn the plentiful deuterium atoms of the world's oceans. In principle therefore, either of these reactor concepts could supply nuclear energy for the long term, and be free, essentially, of nature's resource limitations.

While both the fast-breeder and fusion reactors represent major research and developmental commitments in several countries, the precarious development of the former and the uncertainty about the eventual impact of the latter has provided the impetus to explore additional nuclear energy options and developmental strategies. Specific systems concepts variously labeled hybrid reactors, accelerator breeders, fusion-fission systems, symbionts and others in recent years have been investigated with increasing interest. By all indications, these alternatives do display a substantial degree of robustness and viability and hence deserve a more broadly-based scientific-technical emphasis.

As a class, we identify these emerging concepts as *synergetic nuclear energy systems*. By this term, we define a configuration of system components that maintains certain nuclear processes so as to enhance overall nuclear power production and performance by the integrated utilization of complementary nuclear reactions. The words "synergetic" and "synergism" are of Greek origin and imply the cooperative and beneficial working together of various processes and systems elements.

Interestingly, the technological and conceptual foundation for these systems concepts rests essentially on the integrated use of existing fission reactors and the application of technical advances attained in fast-reactor, fusion reactor and accelerator technology. The synergetic point of view suggests that an integration of the various processes will allow the establishment of a viable nuclear power system while concurrently relaxing the more stringent technical requirements demanded by the independent implementation of separate fast-breeders and fusion reactors. The reasons for this are attributable to compensatory effects associated with energy and isotopic exchanges internally feasible in a synergetic system. We attempt a graphical depiction of this nuclear energy synergetic point of view in Figure 1, using the well-known curve-of-binding-energy as a basis.

Although the various nuclear energy synergetic concepts still are in an embrionic stage, sufficient conceptual and design work already has been reported to indicate some definite features of these systems. The most important of these is the fuller utilization of the more plentiful fertile nuclear materials and an expanded option of nuclear fuel cycles.

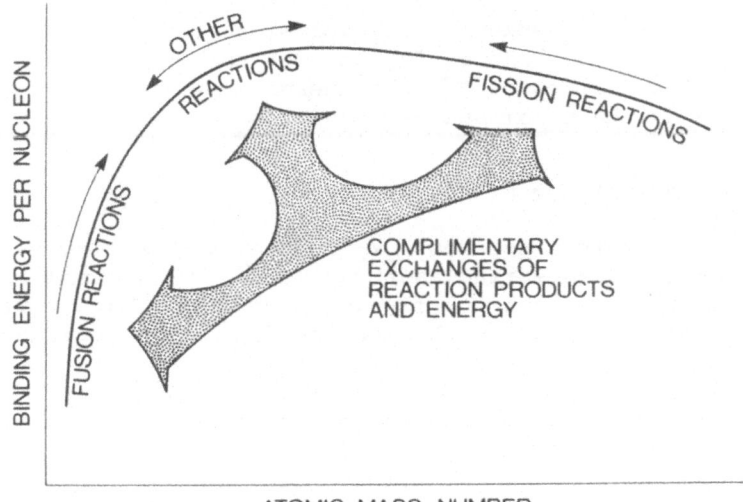

Figure 1.   Schematic Depiction Illustrating Complementary
             Usages of Reaction Products and Nuclear Energy
             Applicable to the Concept of Nuclear Energy
             Synergetics

    Significant benefits apparently can be realized also in
the area of alternate nuclear waste management schemes and
in a reduction of nuclear proliferation hazards. Although
these synergetic nuclear systems possess considerable appeal,
it is apparent that substantial scientific/technical and
institutional/policy reorientation will be needed to bring
these concepts to fruition.

    While the literature on these synergetic systems has be-
come extensive particularly in the area of promotion of
specific proposals it is our purpose here to establish a
systematic and conceptual framework that may provide a broad
perspective of the entire domain of synergetic nuclear energy
concepts. It is expected that subsequent issues of this journal will carry articles on specific systems, design features
and performance parameters.

II. HISTORICAL DEVELOPMENT

The historical evolution of nuclear energy synergism has its roots in the pioneering days of the nuclear energy program. The fission of ^{235}U by thermal neutrons was early identified as the vital nuclear reaction:

$$^{235}U + n \rightarrow {}_5\nu_f n + FP + (\sim 200 \text{ MeV}). \tag{1}$$

Here, n represents a neutron, ${}_5\nu_f$ is the average number of fission neutrons appearing as a result of ^{235}U thermal fission, and FP represents the fission products. Three pertinent features associated with Equation (1) were recognized early:

a.   the reaction possesses an extremely large energy gain;

b.   the terrestrial supply of ^{235}U is highly limited in relation to the potential future energy demand;

c.   the technology for utilizing this fission process was at hand.

The potential fissile fuel exhaustion limitation is, of course, the principle motivation for the development of the fast-breeder and fusion energy. This not withstanding, the fact that the thermal fission technology which by now has been well established, and that the thermal fission reaction is energetically extremely favorable suggests its utilization to the largest extent possible. The question of how to circumvent the ^{235}U exhaustion concerned some of the early nuclear energy scientists. The breeding of fissile ^{233}U and ^{239}Pu induced by neutron capture in the more plentiful ^{232}Th and ^{238}U seemed the only alternative; these breeding chains are given by

$$\begin{Bmatrix} ^{232}Th \\ ^{238}U \end{Bmatrix} + n \rightarrow \begin{Bmatrix} ^{233}Th \\ ^{239}U \end{Bmatrix} \rightarrow \begin{Bmatrix} ^{233}Pa \\ ^{239}Np \end{Bmatrix} \rightarrow \begin{Bmatrix} ^{233}U \\ ^{239}Pu \end{Bmatrix} \tag{2}$$

The decay half-lives were known to be in the range of minutes to days and thus were quite appropriate. What appeared a serious problem, however, was the availability of neutrons

to initiate the foregoing transmutation chain. While it
was known that $_5\nu_f$ in Equation (1) was sufficient to maintain the ^{235}U chain reaction, it was recognized that once
suitable allowances for neutron losses due to escape and
parasitic removals were made, one could replace only a
fraction of the consumed fissile nuclei. Early investigators then searched for non-fission neutron sources. It
is in this attempt that the early ideas of nuclear synergism
emerged.

Documents now declassified provide us with the interesting and significant searches for non-fission neutron sources.
The earliest pertinent record available to this writer is
an in-house report dated 25 August 1952, and authored by
W. B. Lewis of the Chalk River Nuclear Laboratories[1]. In
this paper, Lewis discusses certain published experimental
and theoretical results dealing with neutron emission by
heavy nuclei when excited to high energies and recognizes
the potential of breeding fissile material in an accelerator target; the direct quotation of 1952 is remarkably
contemporary:

"The significance of this phenomenon (i.e., neutron
emission in excited nuclei) is that the high yield
of neutrons makes possible in theory a cyclic reaction in which electrical power is applied to produce high energy excitation of such heavy nuclei
giving a high yield of neutrons. These in turn
are applied to produce fissionable isotopes. By
the chain fission of these isotopes, more electrical
energy is generated than was initially supplied.
This is a cycle giving a net gain of power and
not depending on the natural occurrence on earth of
^{235}U, or other naturally fissionable substance."

The concept suggested by Lewis is what is now variously
called electronuclear breeding, the accelerator breeder,
the spallation symbiont and other related labels. We will
refer to this concept again.

A year later, in 1953, the California Research Development Company[2] submitted a proposal for what was called a
"Driven Thermonuclear Reactor". Here it was proposed to use
a preionized plasma as a target for light ions such as
deuterons or tritons. The interesting feature was the

suggestion to surround the plasma chamber with depleted
uranium in order to realize neutron multiplication by fast
fission induced by the 14 MeV D-T fusion neutron, and
thereby concurrently breed ^{239}Pu and tritium in ^{238}U and
^{6}Li, respectively. Presently well-known nuclear energy
scientists such as E. Teller and R. Post are listed as
contributors to this pioneering program.

The next independent contribution appears to have been
made by J. D. Lawson of England in 1955 (his name currently
is widely used in connection with the Lawson Criterion in
fusion research). In a Harwell document(3) dated December,
1955, Lawson discusses a number of novel D-T fusion possibilities and proceeds to discuss what he calls a "Fusion-
Fission System". The essential point made by Lawson is
that, regardless of how D-T fusion is attained, the attendant 14 MeV fusion neutrons are to strike a surrounding
blanket of ^{238}U, leading to an increase of energy from 22
MeV to 122 MeV per D-T fusion reaction; the concept of energy
multiplication was thus introduced.

The above three references illustrate the initiating
development of nuclear energy synergetic ideas. While isolated studies continued to appear in the literature, the
subject did not arouse much interest until 1969. In that
year, two papers appeared that provided a most pertinent
and apparently realizable illustration of synergetic concepts in terms of current technology. In June 1969, a letter
by Jung(4) contained the derivation of approximate equations
for fusion-fission reactors and fusion-fission-tube reactors
while in November of the same year, Lidsky(5) discussed a
symbiont fusion-fission system; that is, a fusion plant
physically separate from a companion fission reactor but
joined by isotope couplings.

Since the publications by Jung and by Lidsky, numerous
papers have appeared in the open literature. Indicative of
research activity in this field is the number of pertinent
papers read at national(6) and international conferences(7).
One bibliographic assembly(8) contains in excess of 150
recent publications on hybrid concepts alone. An indication
of national interest in these concepts is suggested by the
technical information exchanges between the United States
of America and the Union of Socialist Soviet Republic(9) and
the formal agreement on a specific developmental project(10).

III. SYSTEM NOTATION

Referring to Figure 1, it is obvious that the inclusion of all pertinent nuclear reactions in a synergetic nuclear energy system description would add considerable notational complexity. This would severely detract from our attempt to classify and analyze. Hence, we will define a short-hand notation that appears well suited for our present purposes. One important characteristic feature of this notation is that it emphasizes only the dominant energy producing and isotope breeding reactions. We illustrate this shorthand notation by reference to well-known nuclear energy concepts.

The basic ^{235}U thermal fission process, Equation (1), can be written in the form

$$[^{235}U] \rightarrow {}^{235}U(n,f)_5\nu_f n + (\sim 200 \text{ MeV}) \tag{3}$$

Most of the symbols are used as previously defined. The square brackets are used to indicate that the contained isotope is supplied from external sources, while the curved connecting arrow indicates that one of the fission neutrons is used to induce fission.

As previously indicated, a fundamental feature of synergism is the expanded utilization of fertile nuclei. Using Fe as the symbol for the two fertile nuclei of Equation (2), $Fe = \{^{232}Th, {}^{238}U\}$, the breeding of fissile nuclei $Fi = \{^{233}U, {}^{238}U\}$, may be compactly written as

$$[Fe] \rightarrow Fe(n,\gamma)Fi \tag{4}$$

where γ is used to indicate that various radiations are emitted. Equation (3) hence may be written as

$$[Fi] \rightarrow Fi(n,f)\nu_i n \rightarrow E \tag{5}$$

where $Fi = \{^{233}U, {}^{235}U, {}^{239}Pu\}$, E is the fission energy released per reaction, and ν_i represents the corresponding average number of fusion neutrons emitted.

This shorthand notation can be conveniently used to describe various fission and fusion reactors. An obvious restriction in its use is that it implies that equilibrium fuel cycles have been established.

A. The Fission Burner

The fission burner, invariably associated with highly-enriched fissile fuel, is simply represented in Figure 2. In this representation, we use a rectangular box to convey the meaning of a specific physical facility; the symbol +P is used to suggest that net power is produced. Though normally, Fi represents ^{235}U, clearly, Fi can represent any of the three fissile nuclei ^{233}U, ^{235}U or ^{239}Pu.

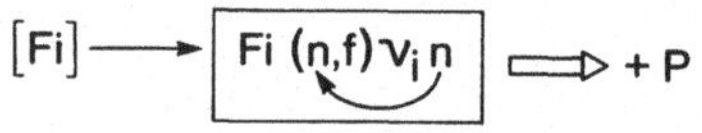

$$[Fi] \longrightarrow \boxed{Fi\ (n,f)\,\nu_i\,n} \Longrightarrow +P$$

Figure 2. Isotope and Reaction Schematic of the Fission Burner

B. The Fast Breeder

The fast breeder incorporates both fissile fuel breeding and fissile fuel burning. Using both Equation (4) and Equation (5), this reactor may be represented as illustrated in Figure 3. The most important feature to emphasize here is, of course, the production of fissile fuel for an external market. As mentioned previously, both Fe and Fi can refer to any one of several specific isotopes.

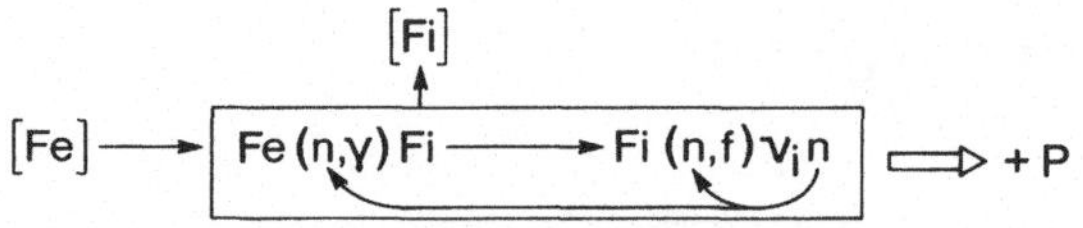

$$[Fe] \longrightarrow \boxed{\begin{array}{c} [Fi] \\ \downarrow \\ Fe\,(n,\gamma)\,Fi \longrightarrow Fi\,(n,f)\,\nu_i\,n \end{array}} \Longrightarrow +P$$

Figure 3. Isotope and Reaction Schematic of the Fast Breeder

C. The Fission Converter

The fission converter reactor is characterized by the flow of both fertile and fissile materials into the reactor as well as concurrent breeding and burning of fissile nuclides, as shown in Figure 4.

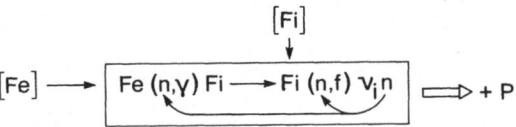

Figure 4. Isotope and Reaction Schematic of the Fission Converter

The significant distinction between the converter and the fast breeder is clearly given by the direction of flow of fissile fuel; the fast breeder supplies an external market with fissile fuel while the converter, as well as the burner, takes [Fi] from an external supply.

We emphasize here that in our chosen notation, fuel enrichment, if any, is not specifically noted; nor are specific reactor parameters such as breeding ratios and burnup specified. These factors are obviously related to specific reactor types and represent a level of emphasis not necessary for our present purpose.

D. The Self-Sufficient Converter

To complete our notation on fission reactors, we introduce the self-sufficient converter as a simple extension of Figures 3 and 4. We impose the requirement that the net flow of fissile isotope to or from external stockpiles be zero, as shown in Figure 5.

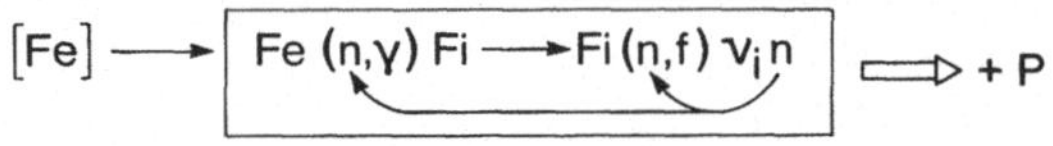

$$[Fe] \longrightarrow \boxed{Fe\ (n,\gamma)\ Fi \longrightarrow Fi\ (n,f)\ \nu_i n} \implies + P$$

Figure 5. Isotope and Reaction Schematic of the Self-Sufficient Converter

It is useful to observe that the notation introduced here
can be used conveniently to define a single parameter that
distinguishes the four fission reactor concepts described.
Recalling the common usage of a fission reactor conversion
ratio, we deduce that for the above fission reactors, the
following holds:

$$
\frac{Fe(m,\gamma)Fi}{Fi(m,f)\nu_i m + Fi(m,\gamma)}
\begin{cases}
= 0, & \text{burner reactor} \\
< 1, & \text{converter reactor} \\
= 1, & \text{self-sufficient reactor} \\
> 1, & \text{breeder reactor}
\end{cases}
\tag{6}
$$

We emphasize again that this applies, in its correct meaning,
only to equilibrium fuel cycle states.

E.   The D-T Fusion Reactor

The first generation fusion reactor is expected to be
based on the $D(T,\alpha)n$ reaction. Almost all fusion reactors
call for a lithium containing blanket around the plasma so
that the tritium required for the fusion reaction is bred
via $^6Li(n,a)T$ and $^7Li(n,n\alpha)T$. In terms of our notation used
here, we write this system concept as illustrated in Figure 6.
It is significant to note that Li and D play a role essentially analogous to Fe and Fi in the fission reactors shown
in Figures 2 to 5.

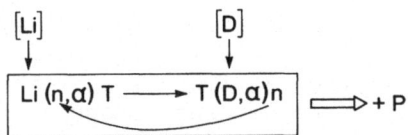

Figure 6.   Isotope and Reaction Schematic of the
D-T Fusion Reactor

F.   The D-D Fusion Reactor

The fusion of two deuterium nuclei is known to occur in

two channels of equal probability

$$D + D \rightarrow {}^{3}He + n$$

$$D + D \rightarrow T + p \tag{7}$$

$$4D \rightarrow {}^{3}He + T + p + n$$

Tritium, and possibly ${}^{3}He$, may further undergo fusion with deuterium yielding

$$D + T \rightarrow n + \alpha$$

$$D + {}^{3}He \rightarrow \alpha + p \tag{8}$$

In general, therefore, we can write

$$XD \rightarrow RP + xn \tag{9}$$

where RP are various reaction products, while X and x are integers. To mold this fusion reaction into our fission notation; e.g., Equation (5), we will write the general D-D fusion reaction as

$$[2D] \rightarrow D(D,RP)\nu_u n \rightarrow E \tag{10}$$

where $\nu_u = x/2X$; thus, ν_u is the fusion analog to the fission ν_i.

With this notation therefore, the D-D fusion reactor is represented as shown in Figure 7.

It is, of course, possible to similarly define numerous other fusion systems simply by drawing upon the specific and relevant reactions. For our purpose here, we restrict ourselves to the above-listed systems as they appear to be primarily relevant in the intermediate term.

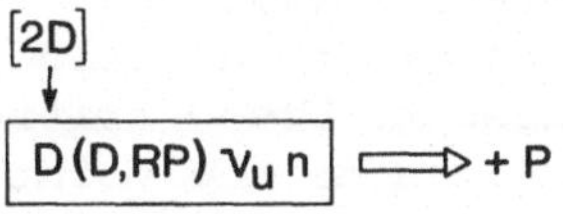

$$[2D]$$
$$\downarrow$$
$$\boxed{D(D,RP)\,\nu_u\,n} \Longrightarrow + P$$

Figure 7.   Isotope and Reaction Schematic of the
D-D Fusion Reactor

IV. SYNERGETIC CONCEPTS

Before we identify and discuss specific energetics nuclear energy concepts, it is useful to focus upon two important considerations: one is neutronic, while the other is
linguistic.

The converter and fast-breeder reactors are characterized by a certain degree of neutron efficiency that is
stressed in system design so as to maximize the replenish-
ment of scarce fissile materials by neutron-induced breeding
via fertile nuclei, as illustrated in Equation (2). An
analogous neutron economy needs to be exercised with the
D-T fusion reactors. This vital role of the neutrons was,
as we indicated previously, a basic consideration during the
initial emergence of synergetic concepts. The contemporary
focus of nuclear energy synergetic is directed not only to
neutron economy but also toward neutron multiplication.

The prospect of utilizing neutron multiplication is
rendered practical by the observation that the energy of
fusion neutrons exceeds that of fission neutrons and that
these energies are in the domain of (n,xn) reaction threshold, while the fission neutrons possess a most probable energy of $\sim$0.8 MeV, the D-D fusion neutrons have an energy from
2.4 MeV, and the D-T neutron, 14.1 MeV. In addition to
threshold (n,xn) reactions, it is known that the neutron
yield increases significantly in the 1 to 14 MeV neutron
energy domain. We illustrate the pertinent data in Figure 8.

An important energy-related consequence follows. Referring to the need to breed tritium in the D-T fusion reactor
as shown in Figure 6, it is common practice to mix various
neutron multiplying materials with lithium. Then each fusion
neutron will lead to one triton, with some fractional or
integral number of neutrons available to breed fissile fuel
in situ, as illustrated in Equations (2) and (4). Since
each fissile nucleus will lead eventually to some 200 MeV,
as illustrated in Equation (1), the extent to which fusion
energy is multiplied depends entirely upon the neutron excess over that required to breed tritium. These considerations can be generalized to the point that the process that
produces neutrons need not, of itself, be energetically self-
sufficient; some of the fission energy can be used to support
this neutron-producing reaction.

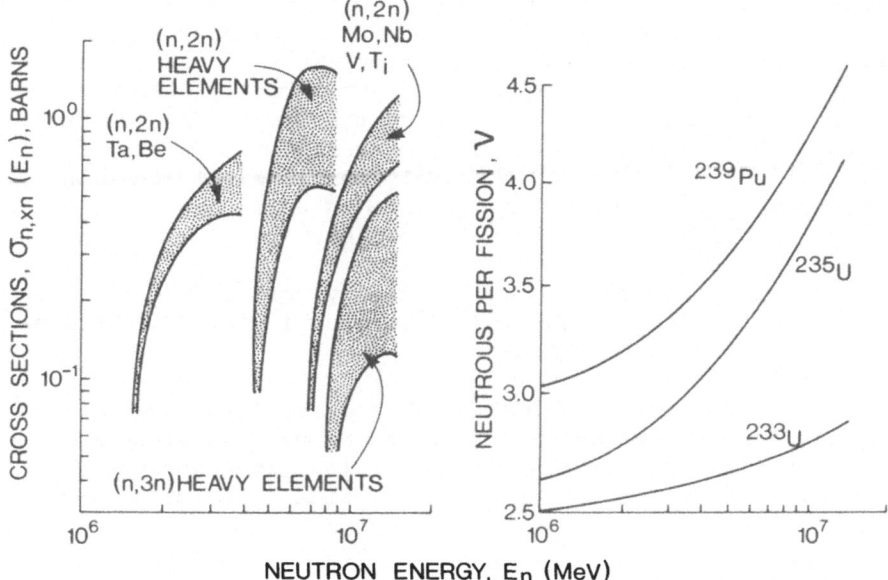

Figure 8.   Approximate Representations of (n,xn) Cross-
Sections and Neutron Yield-per-Fission Pertinent
to Neutron Multiplication Effects up to 14 MeV

It is these considerations that have led to the expression that "fusion is neutron rich and energy poor" while
"fission is energy rich and neutron poor". A complementar-
ity therefore exists, and it follows that if these two
processes could be optimally combined, then advantages might
accrue that could not be realized if such systems were to
operate independently.

This latter view clearly points in the direction of
matching and mixing various neutron producing and energy-
producing reactions within the context of specifically-
designed system configurations. The matter of choosing
names and labels then can become important if ambiguity and
confusion are to be avoided or at least minimized. We elect
here, therefore, to adopt a terminology first introduced in
this context by Lidsky(11). The words "hybrids" and "symbionts", which are of Greek origin(12), will play central

roles. The word "hybrid" conveys the meaning of one organ-
ism of "mixed origin"; here, this will be taken to mean one
reactor based on both fusion, fission and associated reactions. A "symbiont", on the other hand, conveys the meaning
of two dissimilar organisms "living together"; for our purpose, this will be taken to represent separate reactor facilities which, though based primarily on different reaction
processes, will nevertheless operate in an integrated manner by a complémentary flow of isotopes.

We interject here to note that many workers use the word
"symbiote" instead of the word "symbiont"; Klein(12) implies
that the latter is correct.

In the following, we establish a classification of synergetic concepts based on the reactions used to produce the
additional neutrons. Three types of reactions currently
dominate work on the subject: D-T fusion, D-D fusion and
spallation.

A.   D-T Based Synergism

The synergetic systems that have been studied most extensively combine or integrate D-T fusion reactions with fissile-
breeding processes, and in most cases, fission reactions;
fusion of deuterium and tritium can be induced by either mag-
netic or inertial confinement based on essentially any of the
current fusion reactor types.

A broad range of D-T-based fusion-fission conceptual
approaches to the subject are described in references 5,7,
13-30, while the Leonard Bibliography(8) contains numerous
references to specific designs; short summaries as well as
detailed discussion on various proposals are found in references 6 and 9.

Although a number of workers in the field use descriptive
labels such as hybrids, breeders and symbionts interchange-
ably, it appears that in terms of the basic meaning of these
words and emerging usage, three distinct D-T-based fusion-
fission systems can, in principle, be identified; actual designs would clearly represent considerable overlap.

The D-T Hybrid. In its simplest description, the D-T
hybrid consists of a fusion chamber surrounded by a blanket

that provides for (a) breeding of tritium; (b) breeding of
fissile fuel; and (c) fission reactions.

We provide a schematic depiction of such a hybrid with
particular emphasis on the blanket reaction dynamics in
Figure 9. In terms of our shorthand notation, the D-T hybrid
combines fusion, fission and breeding reactions and can be depicted as shown in Figure 10.

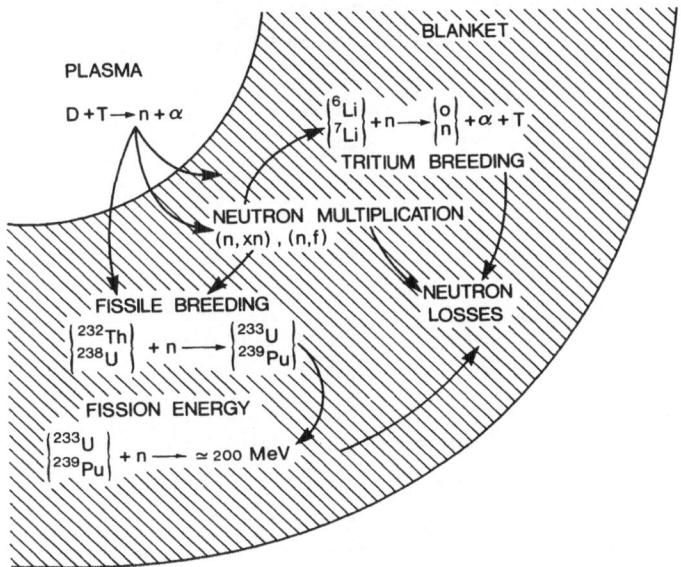

Figure 9.   Illustration of Some Blanket Physics Processes

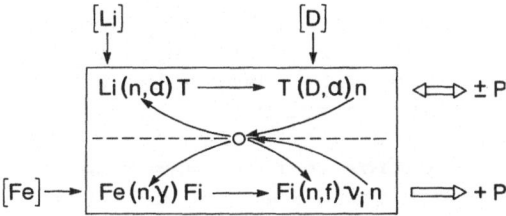

Figure 10.   Isotope and Reaction Schematic of the D-T Hybrid

Clearly, the neutron and isotope flow in the blanket represents considerable complexity; to this, of course, need to be added conventional fission reactor considerations such as heat removal and control. As suggested in Figure 11, the power flow from the fusion core need not be positive. We will comment on this later.

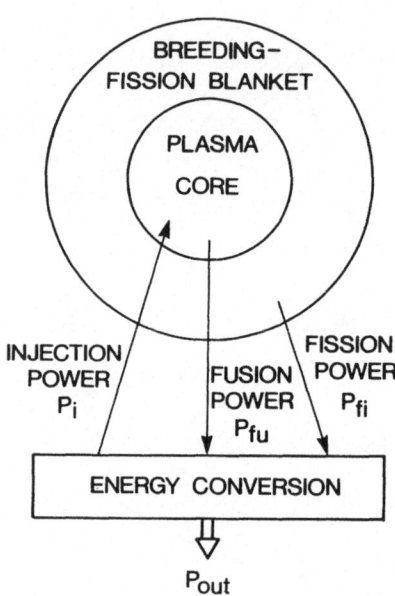

Figure 11.   Simplified Illustration of the Major Components and Principle Power Flow of a Hybrid Reactor

The D-T Breeder.   The D-T breeder, sometimes called a "fuel factory", represents a particular variation or limiting case of the D-T hybrid. Basically, one seeks to minimize the complications attributable to the fission processes in the blanket. This is accomplished by a suitable isotopic throughput control in the blanket so as to minimize fission. The bred fissile fuel in the blanket is used to supply other fission reactors. In terms of our notation, the D-T breeder is described in Figure 12.

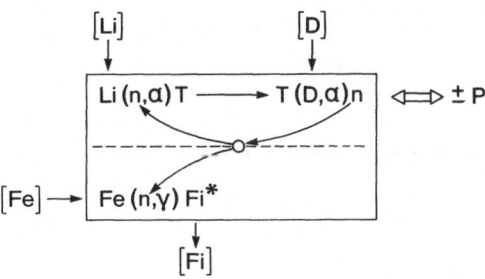

Figure 12.   Isotope and Reaction Schematic of the <u>D-T Breeder</u>

By the use of the asterisk in Fi*, we imply that what is
withdrawn need not be necessarily the final fissile nuclei
^{233}U or ^{239}Pu, but similarly, can be the precursers $\{^{233}$Th,
^{233}Pa$\}$ or $\{^{238}$U, ^{239}Np as suggested in Equation (2). Note,
however, that the tritium breeding function is retained in
full in the blanket. Further, this facility also can be a
net power consumer since it is the eventual use of the bred
fissile fuel that provides energy.

The D-T Symbiont. If the D-T breeder is a basically
simple, though technically significant extension from the
hybrid, the D-T symbiont represents a deliberate and specific
matching between one D-T breeder with one or several companion fission reactors. By combining the D-T breeder, Figure
12, and a fission converter, Figure 4, the D-T symbiont may
be represented as depicted in Figure 13. Although this figure appears very complex, it should be recognized that the
fission component represents existing technology.

As suggested in this form, the high-energy 14 MeV fusion
neutrons are, to some extent, used to breed tritium. However,
as pointed out earlier and illustrated in Figure 9, it is
the higher-energy neutrons that are particularly useful for
the purpose of neutron multiplication purpose, while the
breeding of tritium is best accomplished with thermal neutrons
since the neutron absorption cross-section in ^{6}Li varies as
1/v. Now, since fission reactors normally possess a thermal
neutron spectrum, it may be advantageous to breed tritium in
the companion fission reactor (17,19,29), particularly so if
this would involve neutrons generally lost in flux fattening,
shim and control function purposes.

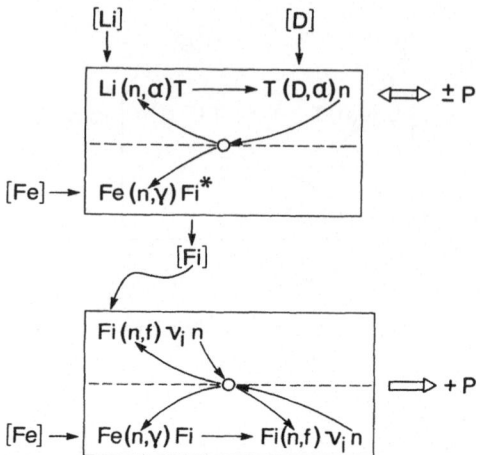

Figure 13.   Isotope and Reaction Schematic of the <u>D-T</u>
             <u>Symbiont</u>

A representation of all possible neutron and isotope
flows is displayed in Figure 14.   The simplified power flow
for this symbiont is depicted in Figure 15 which is essentially equivalent to the hybrid, Figure 11.

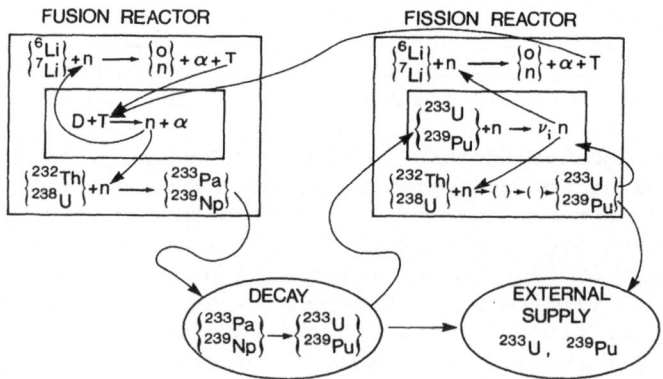

Figure 14.   Isotope Flow for a D-T Fusion-Fission
             Symbiont Reactor System

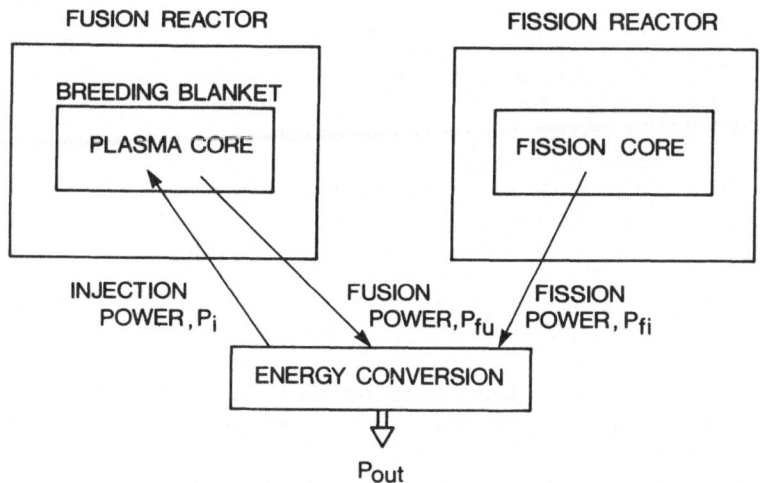

Figure 15.   Schematic of Symbiont Fusion-Fission System

General Comments on D-T-Based Synergetic Concepts.   One
of the points made previously was that in a fusion-fission
system, the fusion component need not be a net power pro-
ducer in order to render the entire concept viable; this can
be demonstrated by reference to Figures 11 and 15.   Using
the notation suggested in these figures, we write the station
electrical power output P_{out} in simplified form as

$$P_{out} = \eta_{fi}P_{fi} + \eta_{fu}P_{fu} - P_i \tag{11}$$

where the η's are the respective thermal-to-electrical conversion efficiencies.   We then use the plasma break-even
parameter, Q_p, defined by

$$Q_p = \frac{\eta_{fu}P_{fu}}{P_i} \tag{12}$$

Substituting then for P_i in Equation (11) yields

$$P_{out} = \eta_{fi}P_{fi} + \eta_{fu}(1-1/Q_p)P_{fu} \qquad (13)$$

where Q_p appears as a term affecting the efficiency of the fusion power contribution to the station electrical output. Interestingly, while for a pure fusion, reactor Q_p must of necessity exceed unity, this restriction can be relaxed within the context of nuclear energy synergism. Using $P_{out} > 0$ as the essential limiting criterion, Q_p obviously must satisfy the inequality

$$Q_p > [1 + (\eta_{fi}/\eta_{fu})(P_{fi}/P_{fu})]^{-1} \qquad (14)$$

Though a more detailed power balance will reveal additional terms in the equations, it will not affect the general conclusion that, in this nuclear synergetic scheme, the price for overcoming the fissile fuel exhaustion limitation with a $Q_p < 1$ plasma is equivalent to a reduction of the system efficiency. Thus, the system also displays some qualities of robustness.

The methodologies employed in the studies of D-T synergetic systems can be grouped into two categories: those based on detailed neutron transport and isotope depletion calculations of specific designs and those based on parametric analysis of a broad range of systems; the detailed transport calculations invariably provide the systems characteristics for the parametric studies.

While the categorization of the D-T synergetic concepts into hybrids, breeders and symbionts provides a convenient classification, it seems that specific system designs would indeed overlap in certain respects. For example, the total elimination of fissile fuel consumption in a D-T breeder could be, indeed, substantially less than in a D-T hybrid and clearly not be zero. Similarly, a D-T hybrid could be designed to operate as a low performance "fuel factory". Thus, while the limiting cases may be clearly specified, a design choice between power and fuel often may have to be made, in which case the extent to which one or the other predominates may determine the label.

The overall fissile fuel production capacity of the hybrid and symbiont is strongly dependent on the characteristics of the blanket and the companion fission reactor (in the case of the symbiont); invariably high internal fission conversion ratios and low specific fissile fuel inventories enhance breeding performance.

As a general characterization of these D-T driven concepts, it is apparent that the fissile fuel breeding capacity is far in excess of that reported for fast breeders; for example, while the fissile fuel doubling time of a fast breeder is measured in terms of decades, a D-T synergetic system can, depending upon the fusion-to-fission power ratios and blanket performance, be of the order of months. Thus, in effect, one D-T synergetic system could, in principle provide fissile material for 5-20 fission converter reactors.

Another way of emphasizing some specific features of these synergetic concepts is to focus upon some specific features of the blanket. Depending upon the details of the blanket design, the plasma fusion energy can be multiplied effectively by factors up to 40, and in some cases, even higher. Similarly, the total number of tritons and fertile atoms bred per fusion neutron is invariably in excess of 2, and may even reach values of 4 to 5.

B. D-D-Based Synergism

The motivation to consider D-D fusion as a component in a nuclear energy synergetic system is essentially equivalent to the reasons put forth for its eventual replacement of the D-T fusion process: (a) reduction of tritium radiological hazards; (b) the elimination of lithium as a potential resource limitation; and (c) simplified blanket design. Of particular relevance is, again, the previously-stated point that energy break-even of plasma is not essential in order to maintain an apparently viable system.

In comparison to D-T schemes, the number of full journal papers on D-D synergetic concepts is sparse(15,31-36); however, short communications and specific proposals have been presented at specialized conferences(6). Based on these, it is apparent that the classification of D-D synergetic systems is essentially analogous to the D-T versions.

The D-D Hybrid. As the name implies, the D-D hybrid
consists of a D-D plasma surrounded by a breeding-fissioning
blanket. The fusion neutrons in Equation (10) enter the
blanket and lead to fissile fuel breeding as well as to
fission. The neutronics and isotope flow are suggested in
Figure 16. Comparing the D-D hybrid to the D-T hybrid as
illustrated in Figure 10 shows that the fundamental difference is in the elimination of a lithium feed and tritium
breeding.

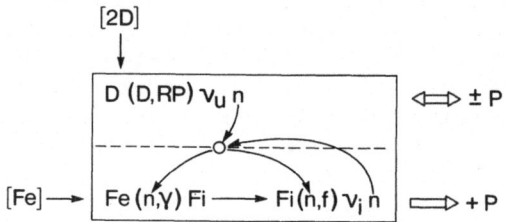

Figure 16.   Isotope and Reaction Schematic of the D-D Hybrid

The D-D Breeder. The D-D breeder is a hybrid with a
substantial, or near total, elimination of fission energy
production in the blanket. This fuel factory is depicted in
Figure 17.

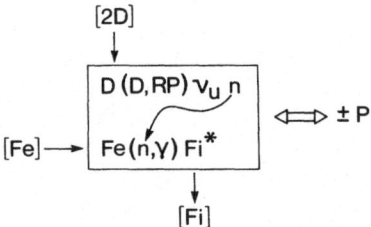

Figure 17.   Isotope and Reaction Schematic of the D-D Breeder

The D-D Symbiont. Integrating the isotopic flow of a
D-D breeder with a fission reactor leads to a D-D symbiont
as shown in Figure 18. Comparing this D-D concept with its

corresponding D-T symbiont as shown in Figure 13 indicates the comparative simplification of the isotopic flow; this difference is ilustrated further by comparing Figure 19 with Figure 14.

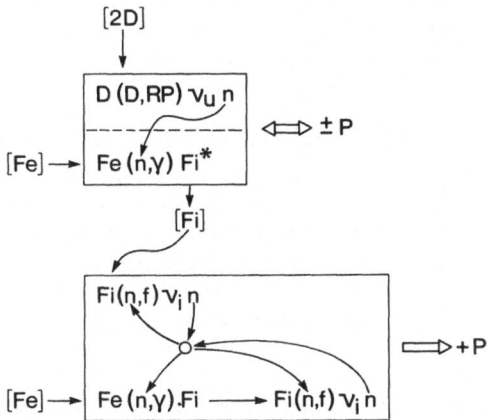

Figure 18.   Isotope and Reaction Schematic of the D-D Symbiont

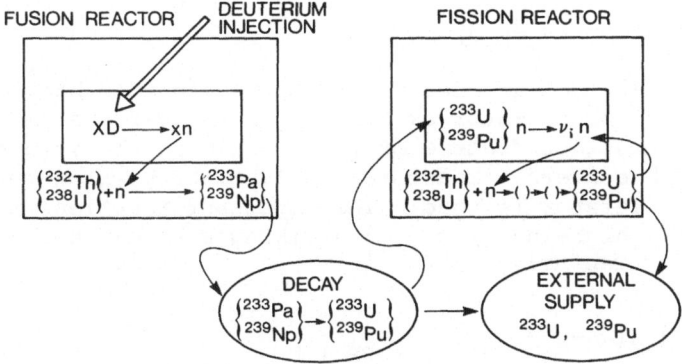

Figure 19.   Isotope Flow for a Beam-Driven D-D
             Fusion-Fission Symbiont

General Comments on D-D Synergetic Concepts. The point
is repeatedly made that the D-D fusion process "avoids" hazards associated with tritium. This is not correct. What is
meant specifically is that while D-D fusion does not require
the breeding and associated handling of tritium, nevertheless there will still exist a tritium inventory due to its
appearance as a fusion reaction product as shown in Equation (7). Under appropriate plasma conditions, this tritium
also will be consumed by fusion as shown in Equation (8).

As a general point, the D-D fusion cycle will be associated with a lower power density requiring, therefore, a larger fusion reactor for a given power output. Now a larger
fraction of the total fusion energy appears in the form of
charged particles and may render direct conversion approp-
riate(37). While the smaller average neutron energies will
have the effect of reducing first-wall damage, when compared
to D-T fusion, it also will decrease the potential neutron
multiplication effect.

In addition to the above, the general comments made
concerning D-T-driven fission reactors also apply to those
based on D-D fusion.

To amplify Figure 19, it appears to be a general view
that neutral beam injectors(38) could be most instrumental
to the attainment of a viable D-D synergetic system.

C.   Spallation-Based Synergism

The next nuclear energy concept we consider here may be
labeled the "Accelerator or Spallation Breeder System"(6,39-
47). The distinct feature of this system is that, in assoc-
iation with converter reactors, a high-energy, high-current
proton accelerator is used. The important reaction is that
involving a high-energy proton striking a high-Z nucleus such
as lead or uranium:

$$p + \left\{ \begin{array}{c} \text{High-Z} \\ \text{Nucleus} \end{array} \right\} \rightarrow \nu_s n + RP + E \tag{15}$$

The number of neutrons thus produced, ν_s, is dependent
on the proton energy, target composition and target size; as
indicated in Figure 20, a 1 GeV proton can generate some 20

to 100 neutrons. These, neutrons appear as a result of
nuclear cascade interactions, neutron evaporation, and fast
fission as shown in Figure 21. The neutron energy spectrum
of the evaporation neutrons is known to be somewhat harder
than a fission neutron spectrum, while the cascade neutrons
possess even much higher energies.

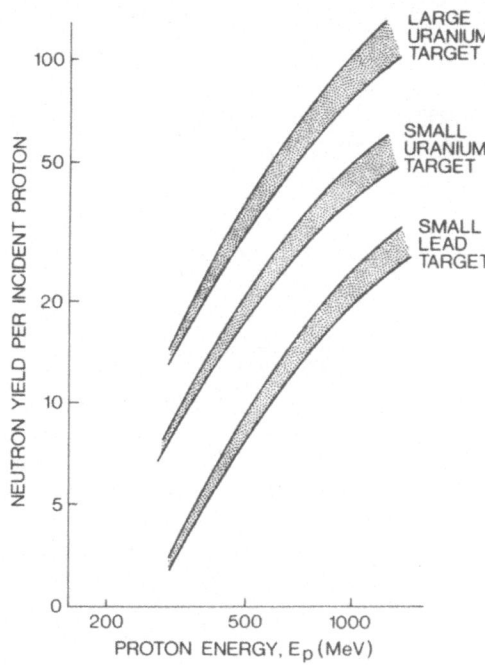

Figure 20.   Neutron Yield in a Spallation Target as a
            Function of Proton Energy and Target
            Size/Compositions

        In terms of our previous symbolic notations, the isotopic
coupling between the accelerator target-blanket arrangement
and the companion fission reactor is depicted in Figure 22,
while Figure 23 provides a more detailed isotopic flow of
the concept.

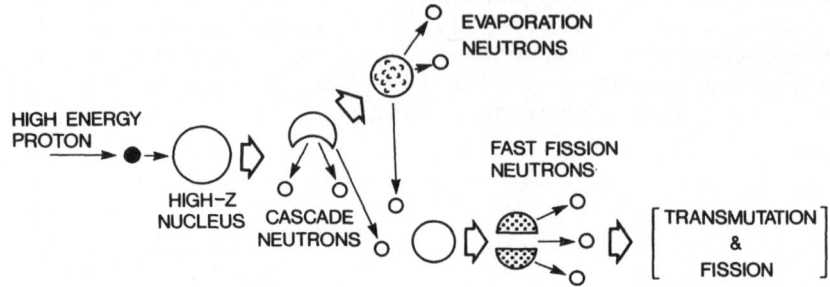

Figure 21. Illustration Showing Possible Processes Induced
by a High-Energy Proton

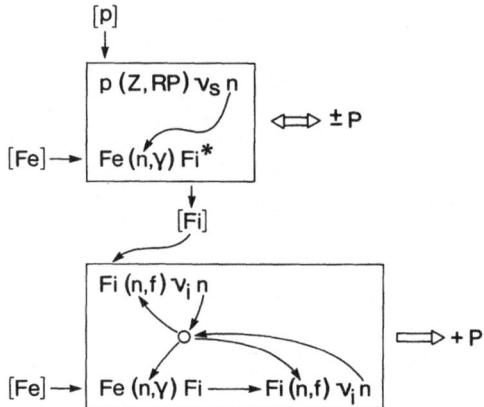

Figure 22. Isotope and Reaction Schematic
of the <u>Spallation Breeder</u>

In addition to the power generated by the companion
fission reactor(s), the accelerator target will contribute
significantly to energy production. Typically, the proton
beam is expected to consist of a 200-300 mA of ∿1000 MeV
protons, thus contributing some 200-300 MW of thermal energy
to the overall system power balance to which secondary reac-

tion energies are added. The target/blanket, therefore, is
a medium-sized power plant and would need to be designed
accordingly. It is useful to note that to a first approximation, the power balance for such a spallation breeder
system would be of the form used for the hybrid as shown in
Figure 11, and the symbiont as shown in Figure 15; the injection of power would refer to the electrical power required
to produce the beam, while the Q value represents the efficiency with which this process can be accomplished.

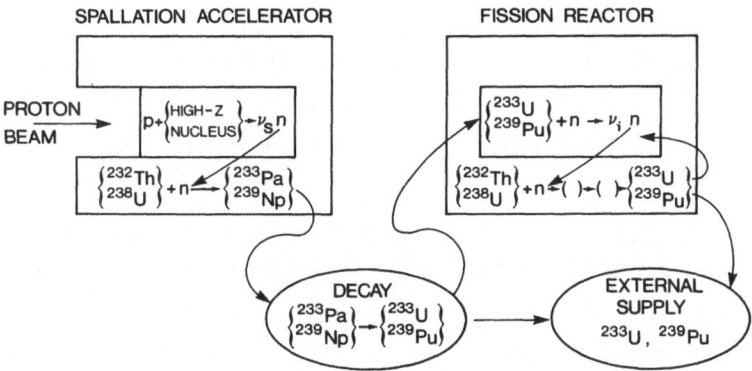

Figure 23.   Schematic Representation of Layout and Isotope
             Flow for a Spallation Breeder System.

As with the fusion hybrid and fusion symbiote, the
combined accelerator and companion fission reactor, when
viewed as a breeder system, can, in principle, also yield
fissile fuel doubling times an order of magnitude shorter
than those of a fast breeder reactor. Experience with low-
current linear accelerates suggests some vital technical
challenges that need to be resolved.

V. NUCLEAR FUEL AND NUCLEAR WASTE MANAGEMENT POSSIBILITIES

In the preceding, we have emphasized the fuel breeding
and energy producing synergetic form of nuclear energy production. In addition to this basic and important feature
of a more enduring energy system, it appears that some interesting emerging options associated with nuclear fuel and

nuclear waste management may be feasible.

One central and distinct feature of each of the several concepts discussed was the useful function of a blanket fed by neutrons whose source energies exceed those produced by fission. The useful nuclear consequences of this combination are based on the following:

a.   The higher neutron energies allow an expanded utilization of threshold reactions (see Figure 8) .

b.   The higher neutron energies combined with the reduced demand for neutron conservation allow more scope for neutron spectrum shaping.

c.   Neutron fluxes (neutron/cm^{2}s) greater than presently available may be achieved.

d.   Additional nuclear reactions achievable with sufficient magnitude and density may be attained.

These considerations may be translated into the following practical consequences:

a.   It is known that the capture-to-fission ratios of pairs of fertile-to-fissile nuclei, that is,

$$\sigma_c(232_{Th})/\sigma_f(233_U) \text{ and } \sigma_c(238_U)/\sigma_f(239_{Pu})$$

attain their maximum in the high keV neutron energy region. This suggests the possibility of in situ enrichment and reenrichment, not only for large central nuclear power stations but also for small, more conveniently situated decentralized nuclear power stations(48-51).

b.   Since the neutron yield in a spallation target increases with proton number, it is suggested that the particularly hazardous actinides accumulated in spent nuclear fuel be used as accelerator target materials; thus, more neutrons are produced for fissile fuel breeding from the more hazardous nuclear wastes.

c.   It appears to this reviewer that the protons
      released in the spallation-cascade reaction
      could be used to transmute neutron-rich
      fission products.

d.   With the higher neutron flux available in
      an accelerator target/blanket arrangement,
      as well as in the first wall region of a
      fusion blanket, it now may be more readily
      possible to transmute a larger number of
      the hazardous fission products and thus
      obtain what Lidsky (11) labels the "Augean"
      system.

e.   The expanded control over feed stock in a
      fusion or spallation blanket would allow
      more control in the development of new and
      more proliferation-resistant fuel cycles.

These various nuclear fuel and nuclear waste management
schemes are schematically suggested in Figure 24.

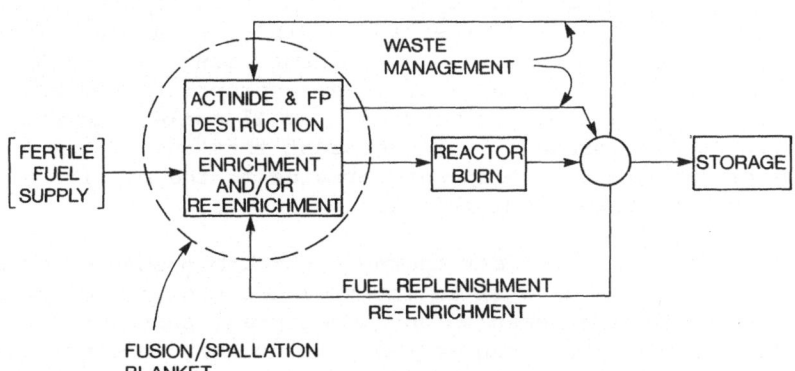

Figure 24.   Illustration of Nuclear Fuel and
              Nuclear Waste Management Possibilities

VI.   CONCLUDING COMMENTS

In the preceding, we have attempted to provide a coherent description of the class of emerging nuclear energy
systems concepts that are based on the complementary utilization and integration of various nuclear processes; that
is, the notion of nuclear energy synergism. By deliberate
intent, our emphasis has been chosen to be analytical and
impressionistic rather than enumerative and encyclopedic.
The reader interested specifically in design details and
evaluated performance parameters will find the cited references a useful resource for information along these lines;
a listing of design proposals, each with its own particular
fuel cycle and performance characteristics, would make a
review here far too tedious and dull, and therefore is not
included.

There are, however, some points not discussed so far in
this review that nonetheless deserve passing mention. The
several technologies discussed, D-T fusion, D-D fusion and
spallation accelerators, represent large-scale, costly under-
takings which in their entirety are simply beyond the scientific and financial resources of most countries. While such
countries nonetheless could make useful scientific-technical
contributions to certain aspects of large-scale nuclear energy
synergism along the above-mentioned lines, it appears that
the deployment of upgraded dense plasma focus devices(52)
could be developed potentially as an appropriate context for
synergetic systems development on a small scale. However,
the subject is pursued, it is apparent that one of the urgent
areas of immediate development pertains to the physics and
technology of the breeding-fission blankets.

The view that nuclear energy synergism possesses considerable robustness as well as interrelatedness with existing
fission and fusion research and development pervades much of
the literature on the subject; a particularly succinct and
direct statement along this line is presented by Augenstein
(53). It follows then that since nuclear synergism possesses many elements of intermediate technology- intermediate
in the sense of being between the proven and established
technology of thermal fission energy on the one hand, and
the uncertain technology and performance of fast breeder and
pure fusion energy on the other- it may very well fit into
the role of providing an assured supply of nuclear ener-

gy in the transition from a thermal fission economy to an
ultimate resource-unlimited nuclear economy more attuned
to a variety of external demands. If nuclear energy synergetic ideas provide a useful role toward this transition,
then they have indeed served well in contributing to a more
acceptable nuclear energy future.

REFERENCES

1.   Lewis, W. B., "The Significance of the Yield of Neutrons
      from Heavy Elements Excited to High Energies," Chalk
      River Nuclear Laboratories, Canada, DR-24, 1952; re-
      printed as AECL-968, 1960.

2.   California Research and Development Company, "Proposal
      for a Driven Thermonuclear Reactor," United States
      Atomic Energy Commission, U. S. A., Report No. LWS-
      24920, 1953.

3.   Lawson, J. D., "A Survey of Some Suggested Methods of
      Realizing Fusion Reactors," Atomic Energy Research
      Establishment, Harwell, United Kingdom, AERE-GP/M,
      No. 185, 1955.

4.   Jung, H., "Two Suggestions Regarding Controlled Fusion:
      Approximation Equations for Fusion-Fission Reactors and
      Fusion-Reaction-Tube Reactors," Nuclear Fusion, 9,
      Page 169, 1969.

5.   Lidsky, L. M., "Fission-Fusion Symbiosis: General
      Considerations and a Specific Example," Proceedings
      Nuclear Fusion Reactors Conference, 17-19 September
      1969, Culham Laboratory, United Kingdom.

6.   See, for example, recent issues of, Transactions of the
      American Nuclear Society, Proceedings of the Topical
      Meetings on the Technology of Controlled Nuclear Fusion
      and similar publications.

7.   International Workshop on Emerging Concepts of Advanced
      Nuclear Energy Systems, 29-31 March 1978, Technical
      University of Graz, Graz, Austria. The papers pre-
      sented at this conference have appeared in two issues
      of Atomkernenergie, 32, 1978.

8. Leonard, B. R., "Bibliography of Fusion-Fission Hybrid Hybrid Publications," Battelle Northwest Laboratories, Richland, Washington, U. S. A., January, 1978.

9. U. S. A.-U. S. S. R. Symposia on Fusion-Fission Reactors held at Livermore, California, U. S. A., 1976, and at Moscow/Leningrad, U. S. S. R., 1977.

10. Gough, W. C., "EPRI/Kurchatov Institute Joint Program on Fusion-Fission," Trans. American Nuclear Society, 27, Page 340, 1977.

11. Lidsky, L. M., "Fission-Fusion Systems: Hybrid, Symbiotic and Augean," Nuclear Fusion, 15, Page 151, 1975.

12. Klein, E., A Comprehensive Entymological Dictionary of the English Language, Elsevier Publishing Co., New York, 1971.

13. Brockman, H., "Neutronic Calculations for a Fusion-Reactor-Blanket Assembly by Discrete Ordinates and Monte-Carlo Methods," Nuclear Fusion, 12, Page 389, 1972.

14. Leonard, B. R., Jr., "A Review of Fusion-Fission (Hybrid) Concepts," Nuclear Technology, 20, Page 161, 1973.

15. Horoshko, R. N., Hurwitz, H., and Zmora, H., "Application of Laser Fusion to the Production of Fissile Materials," Annals Nuclear Science Engineering, 1, Page 233, 1974.

16. Kolesnichenko, Y. I. and Reznik, S. N., "D-T Plasma as a Source of Neutrons for the Combustion of Uranium-238," Nuclear Fusion, 14, Page 114, 1974.

17. Harms, A. A., "Hierarchical Systematics of Fusion-Fission Energy Systems," Nuclear Fusion, 15, Page 939, 1975.

18. Golovin, I. N., "The Position of Hybrid Reactors in Power-Generation Systems," Soviet Atomic Energy, 39, Page 1035, 1975.

19. Gordon, G. W. and Harms, A. A., "Comparative Energetics of Three Fusion-Fission Symbiotic Nuclear Reactor Systems," Nuclear Eng. Design, 34, Page 269, 1975.

20. LaVergne, G., Robinson, J. E. and Martel, J. G., "on the Matching of Fusion Breeders to Heavy-Water Reactors," Nuclear Technology, 26, Page 12, 1975.

21. Maniscalco, J., "Fusion-Fission Hybrid Concepts for Laser Induced Fusion," Nuclear Technology, 28, Page 98, 1976.

22. Harms, A. A., "Upper Bounds of Fissile Fuel Yield with Fusion Breeders," Canadian Journal Physics, 54, Page 1637, 1976.

23. Shang-Fou, S., Woodruff, G. L. and McCormick, N. J., "A High-Gain Fusion-Fission Reactor Producing Uranium-233," Nuclear Technology, 29, Page 392, 1976.

24. Seifritz, W., "The Symbiosis Between Beam-driven Hybrid DT-Fusion Reactors and Near-Breeder HTGR's," Atomwirtschaft, 21, Page 205, 1976.

25. Harms, A. A. and Gordon, C. W., "Fissile Fuel Breeding Potential with Paired Fusion-Fission Reactors," Annals Nuclear Energy, 3, Page 411, 1976.

26. Cook, N. G. and Maniscalco, J. A., "Uranium-233 Breeding and Neutron Multiplying Blankets for Fusion Reactors," Nuclear Technology, 30, Page 5, 1976.

27. Lee, J. D., Editor, "Mirror Hybrid Reactor Studies," Lawrence Livermore Laboratory, U. S. A., UCRL-50043-1,

28. Fortescue, P., "Comparative Breeding Characteristics of Fusion and Fast Reactors," Science, 196, Page 1326, 1977.

29. Blinkin, V. L. and Novikov, V. M., "Symbiotic System of a Fusion and a Fission Reactor with Very Simple Fuel Reprocessing," Nuclear Fusion, 18, Page 7, 1978.

30. Bethe, H. A., "The Fusion Hybrid," Nuclear News, 27, Page 41, May 1978.

31. Powell, C. and Hahn, D. J., "Energy Balance of a Hybrid
     Fusion-Fission Reactor," Atomkernenergie, 21, Page 175,
     1973.

32. Kolesnichenko, Y. I. and Reznik, S. N., "The D-D Nuclear
     Fusion Reaction in a Hybrid Reactor," Nuclear Fusion,
     16, Page 1, 1976.

33. Nakashima, H., Okta, M. and Seki, Y., "Nuclear Charac-
     teristics of D-D Fusion Reactor Blankets (I)," Journal
     Nuclear Science Technology, 14, Page 75, 1977.

34. Greenspan, E., "On the Feasibility of Beam-Driven Semi-
     Catalyzed Deuterium Fusion Neutron Sources for Hybrid
     Reactor Applications," Princeton Plasma Physics Labor-
     atory, Princeton, New Jersey, PPPL-1339, 1977.

35. Schoepf, K. F. and Harms, A. A., "The Synergetics of
     the Catalytic D-D Fusion Breeder," Nuclear Fusion, 19,
     Page 5, 1979.

36. Schoepf, K. F. and Harms, A. A., "Characteristics of a
     Beam-Driven Deuterium Fueled Fusion-Fission Reactor
     System," Nuclear Science Engineering (scheduled for
     August Issue, 1979).

37. Miley, G. H., Fusion Energy Conversion, American Nuclear
     Society, LaGrange Park, Illinois, U. S. A., 1976.

38. Jassby, D. L., "Neutral Beam Driven Tokamak Fusion
     Reactors," Nuclear Fusion, 17, Page 309, 1977.

39. Church, T. G., Editor, The ING Status Report, Chalk
     River Nuclear Laboratories, Canada, AECL-2750, 1967.

40. Davidenko, V. A., "On Electronuclear Breeding," Soviet
     Atomic Energy, 29, Page 866, 1970.

41. Vasilkov, V. G., et al., "The Electronuclear Method of
     Generating Neutrons and Producing Fissionable Materials,"
     Soviet Atomic Energy, 29, Page 858, 1970.

42. Tunnicliffe, P. R., Chidley, B. G. and Fraser, J. S.,
     "High Current Proton Linear Accelerator and Nuclear
     Power," Chalk River Nuclear Laboratories, Canada, AECL-
     5622, 1976.

43. Van Atta, C. M., Lee, J. D. and Heckrotte, W., "The Electronuclear Conversion of Fertile to Fissile Material," Lawrence Livermore Laboratory, U. S. A., UCRL-52144, 1976.

44. Steinberg, M., et al., "Linear Accelerator-Breeder (LAB)," Brookhaven National Laboratory, U. S. A., BNL-50592, 1976.

45. Harms, A. A. and Gordon, C. W., "A Parametric Analysis of the Spallation Breeder," Nuclear Science Engineering, 63, Page 336, 1977.

46. Kouts, H. J. C. and Steinberg, M., Editors, Proceedings Information Meeting on Accelerator-Breeding, Brookhaven National Laboratory, CONF-770107, 1977.

47. Greenspan, E., "Preliminary Report on the Promise of Accelerator-Driven Natural-Uranium-Fueled Light-Water-Moderated Breeding Power Reactors," Oak Ridge National Laboratory, U. S. A., ORNL/TM-6138, 1978.

48. Schultz, K. R., et al., "A U-233 Fusion-Fission Power System Without Reprocessing," General Atomic Company, U. S. A., GA-A14635, 1977.

49. Harms, A. A. and Hartmann, W., "Spent Nuclear Fuel Re-enrichment Without Reprocessing," Annals Nuclear Energy, 5, Page 213, 1978.

50. Grand, P. and Kouts, H. J., "Conceptual Design and Economic Analysis of a Light Water Reactor Fuel Regenerator," Brookhaven National Laboratory, U. S. A., BNL-50838-UC-80, 1978.

51. Jassby, D. L., "Fusion-Supported Decentralized Nuclear Energy System," Princeton Plasma Physics Laboratory, U. S. A., PPPL-TM-316, 1978.

52. Harms, A. A. and Heindler, M., "The Matching of Dense Plasma Focus Devices with Fission Reactors, Nuclear Science Engineering, 66, Page 1, 1978.

53. Augenstein, B, "Fusion-Fission Hybrid Breeders: Economic and Performance Issues, Role of Advanced Converters, Interdependence Between Fission and Fusion Programs," The Rand Corporation, U. S. A., P-6047, 1977.

VAPOR EXPLOSION PHENOMENA

WITH RESPECT TO NUCLEAR REACTOR SAFETY ASSESSMENT

A. W. Cronenberg and R. Benz

Institut für Kernenergetik, Universität Stuttgart

7-Stuttgart-80, West Germany

I.   INTRODUCTION

An important concern in the analysis of a hypothetical nuclear power reactor accident is an understanding of the consequences of reactor core overheating, leading to fuel melting and subsequent interaction of hot molten fuel with coolant. If such molten fuel-coolant interaction (MFCI) is of limited extent, the resultant work potential is relatively benign. However, as illustrated in Figure 1, it can be envisioned that under certain conditions, core overheating may lead to a sequence of events resulting in the formation of an extensive amount of hot molten fuel in a liquid coolant environment, where such molten fuel may interact with the colder liquid coolant, causing it to vaporize as a result of local heat transfer. If the local heat transfer process is rapid enough (for example, due to fine-scale fuel fragmentation and intermixing with the coolant), the vapor generation process may be extremely fast, such that shock pressurization of the system occurs. If the pressure pulse generated is of sufficient strength, then severe damage to or failure of the reactor vessel may occur. Such a process is often referred to as a "vapor explosion".

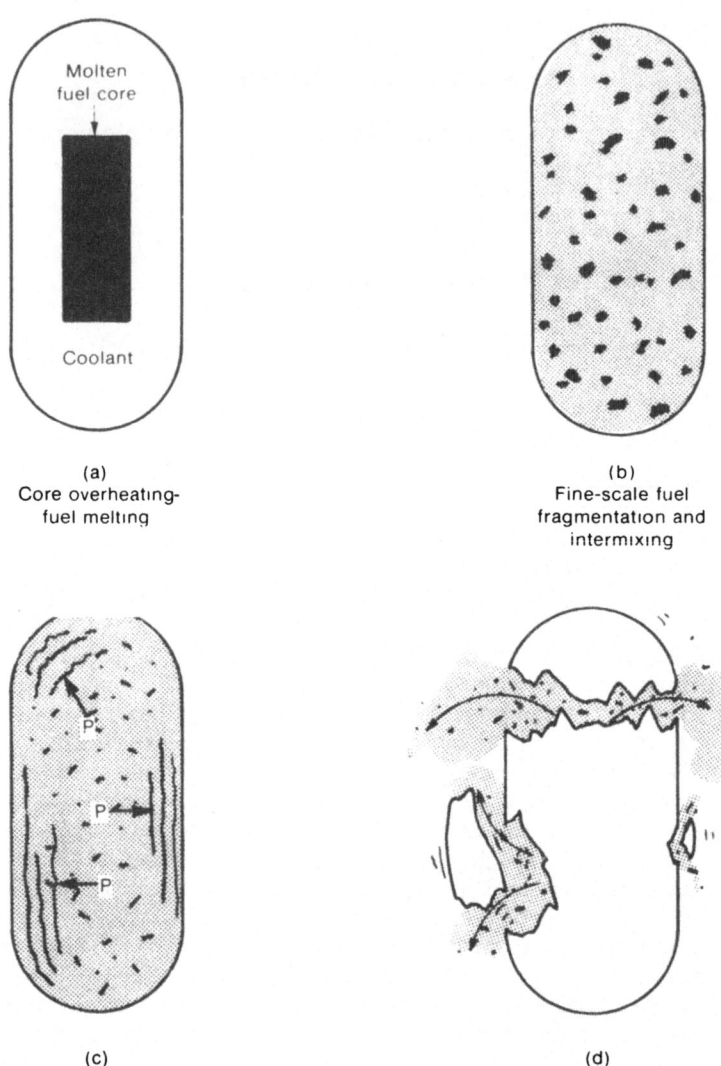

(a)
Core overheating-
fuel melting

(b)
Fine-scale fuel
fragmentation and
intermixing

(c)
Rapid vaporization
and shock pressurization

(d)
Potential reactor
vessel failure

Figure 1.   Illustration of a Hypothetical Molten Fuel-
Coolant Interaction Induced Vapor Explosion

The implication of this type occurence, where the reactor vessel is breached and releases high-temperature material that is radioactive, can be severe. Although such a
vapor explosion accident has never been known to occur in
an operating commercial power reactor, the small possibility of such an occurrence and the stringent safety requirements placed on the licensing of nuclear reactors for commercial use has stimulated research aimed at understanding
the basic phenomena involved in such vapor explosions.

Although the general subject area of MFCI's is of
interest in overall nuclear reactor safety assessment, where,
for example, mild MFCI events can lead to off-normal coolant
hydraulic effects, the present discussion concentrates on
the specific case of severe MFCI-induced explosive vaporization, where a critical review of recent developments in
the understanding of vapor explosion phenomena is presented.
Emphasis is placed on reviewing and assessing the validity
of assumed processes associated with several overall vapor
explosion models, as well as an understanding of fine-scale
fuel fragmentation and intermixing with liquid coolant, which
is considered necessary for the occurrence of large-scale
explosions. In addition, considerations relevant to nuclear
reactor systems are addressed. From such a critical review,
it is hoped that the reader will obtain a better understanding of the major contributions in describing the necessary
conditions for explosive vaporization, the detailed theoretical modeling of such conditions, the validity and limitations of the various proposed models, and the principal uncertainties that still exist at the time of this writing.
First, however, a brief discussion of MFCI incidents (some
of which can be classified as true vapor explosions) is
presented. Following this discussion, various overall vapor
explosion models are reviewed, as well as modeling and experimental efforts on the fragmentation and intermixing processes. The relation of such models to reactor conditions
is also presented, followed by summary and conclusion sections.

II. MOLTEN FUEL-COOLANT INTERACTION INCIDENTS

The general area of molten fuel-coolant interaction has
been the subject of considerable research(1-7). This prob-

lem, however, is not only of interest with respect to nuclear
reactor safety(8) but also to the safety of the foundry(2,9)
and liquified natural gas(10,11) industries. The problem is
not one of assessing the mechanisms involved in such interactions and the associated work potential resulting from the
contact of a hot substance immersed in a colder fluid with
attendant coolant vaporization and pressurization. As illustrated in Figure 2, such metal-coolant interactions can be
considered vapor explosions when the rate of pressure buildup occurs faster than pressure relief in the surrounding
liquid. Such a situation will lead to a shock-type pressure
pulse and the possibility for severe damage to the confining
system, depending upon the rate and extent of volume expansion of the working fluid.

Early industrial experience has shown that vapor explosions often result during the accidental spillage of molten
metals onto damp surfaces. One such incident, described by
Long(2), resulted when molten steel slag was accidentally
spilled into an open trough containing water, causing extensive foundary damage. Epstein(3) documented other examples
of metal-water explosions. In one case, the loading of wet
scrap aluminum into a furnace produced an explosion causing
six deaths, four injuries, and approximately $1 million in
property damage.

As illustrated in Table I, several molten fuel-coolant
interactions (MFCI's) have occurred in test reactor systems,
some of which can be classified as true vapor explosions.
One of the first indications of an MFCI in a nuclear reactor
was a result of the core-meltdown accident of the experimental Canadian NRX test reactor(12) (Table I). During a
low-power experiment, failure occurred in the shutdown-rod
system due to a concurrence of mechanical defects and operating errors. Hurst(13) concluded, in his analysis of this
accident, that the damage to the calandria tubes was a direct cause of either a uranium-steam or uranium-water interaction, which was of a nonchemical nature. Evidence of a
similar type fuel-coolant interaction (FCI) in a fast reactor dates back to the Experimental Breeder Reactor-I (EBR-
I) (U-Zr alloy fuel) meltdown with NaK coolant(14,15). Although no evidence of severe pressurization was found, apparently, coolant boiling resulting from an MFCI played a
major role in damage to the core(15).

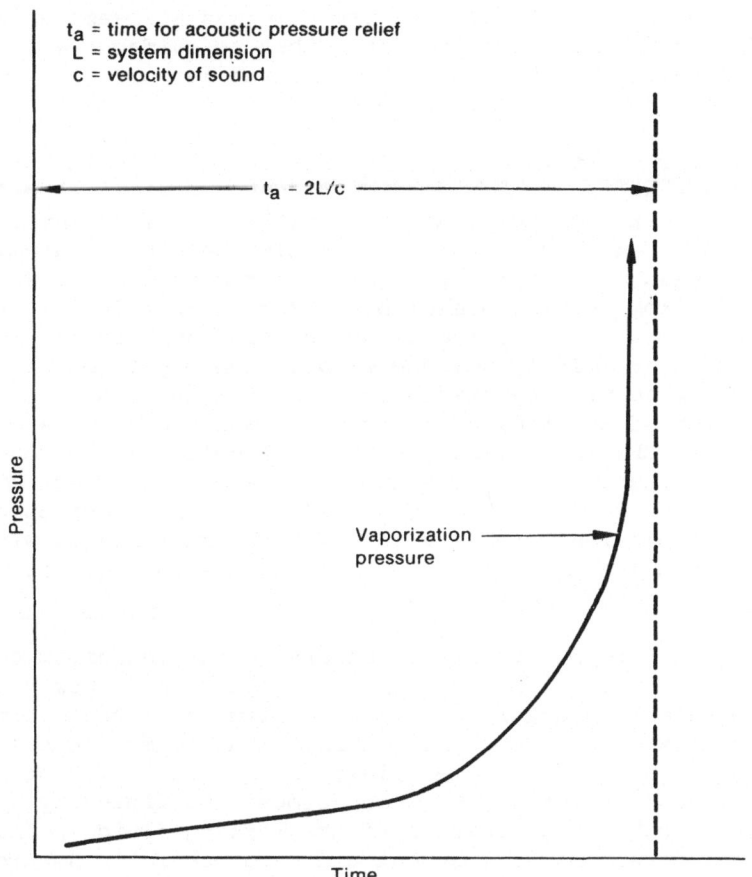

Figure 2.   Transient Pressurization Characteristics Resulting from Explosive Vaporation

Following the NRX incident, a destructive reactor experiment was performed with the Boiling Water Reactor (BORAX)
facility(16,17). As described by Dietrich(16), "A power
excursion melted most of the fuel plates. The pressure resulting from the molten metal in contact with the reactor
water burst the reactor tank and ejected most of the contents
of the shield tank into the air. Analysis of mechanical

damage indicated that the peak pressure was at least as high
as 6,000 psi". Dietrich concluded that no evidence existed
of any significant chemical reaction, but rather that the
explosion was of a thermal nature.

The first fatal reactor accident occurred on January 3,
1961 as a result of the explosion of the experimental military Stationary Low Power Reactor-1 (SL-1) boiling water
reactor(18). Although other causes have been postulated,
the most plausible explanation is that the accident resulted
primarily from the withdrawal of a control rod, leading to
a nuclear excursion, fuel-element failure, and violent interaction between molten fuel and water. The magnitude of the
indicated pressures generated in this accident led to the
development of a series of experiments which concentrated
interest on the mechanisms for MFCI. Results of the controlled destructive experiments with the Special Power Excursion Reactor Test-Idaho (SPERT-ID) core also indicated
that the explosion primarily was of a thermal nature, with
an estimated 10% maximum energy release due to chemical
reaction effects(19-22).

Uncertainty still exists concerning the mechanism of
and factors influencing such thermal explosions. However,
it is generally recognized that extensive fuel fragmentation
and intermixing with coolant occurs, with intimate contact
between the hot and cold materials, resulting in a heat
transfer and coolant vaporization process sufficiently rapid
to cause shock pressurization of the working fluid (coolant).
If such interaction is coherent and large scale in nature,
severe damage to the reactor system can result. Milder fuel-
coolant interactions also can occur, as demonstrated in
numerous in-pile and out-of-pile experiments, that can result in a rather violent vaporization process; however, such
interactions are not considered to be true vapor explosions
unless severe shock-pressurization occurs. Evidence indicates that the SL-1, BORAX-I, and SPERT-ID incidents can be
categorized as undergoing true vapor explosions, whereas the
NRX and EBR-1 incidents can be categorized as being due to
milder MFCI. However, because of redundant safety systems,
such energetic interactions have not occurred in modern
commercial power reactors and their probability of occurrence
is low(8).

TABLE I

SOME WATER REACTOR ACCIDENTS AND EXPERIMENTS INVOLVING METAL-WATER INTERACTION

Date of Occurrence	Reactor	a.Fuel b.Coolant c.Moderator	Fuel-Coolant Geometry	Cause
12/12/52	NRX	a.Natural-U clad with Al b.H2O c.D2O		Control rod maloperation during an experiment on the reactivity of the reactor at low power. The object of the experiment was to compare the reactivity of irradiated fuel with fresh fuel. The normal water coolant flow for some of the rods was reduced for this experiment.

Description of Events	Primary Factors Influencing Reactor System Damage
A postaccident investigation led to the conclusion that the reactor went supercritical by about $0.6 and that the power rose to about 20 MW(t). Although the reactor was capable of operating at power levels up to 30 MW(t), the reduced cooling rate for the test resulted in coolant boiling and steam pressure buildup, causing eventual melting of several rods and subsequent interaction between U and Al with the water coolant.	a.Coolant vaporization leading to fuel rod meltdown b.Steam pressure buildup (non-explosive) c.Hydrogen-oxygen explosion

Analysis of the accident indicated that the exothermic chemical reaction was not significant in itself. However, the attendant hydrogen release from this reaction, such that the core and calandria tubes were damaged beyond repair. There was no evidence of a shock-type steam explosion.

TABLE I

Date of Occurrence	Reactor	a.Fuel b.Coolant c.Moderator	Fuel–Coolant Geometry	Cause
1954	BORAX-l	a.U-Al Alloy clad in in Al b.H₂0 c.H₂0		Destructive experiment initiated by 4%-keff control rod ejection

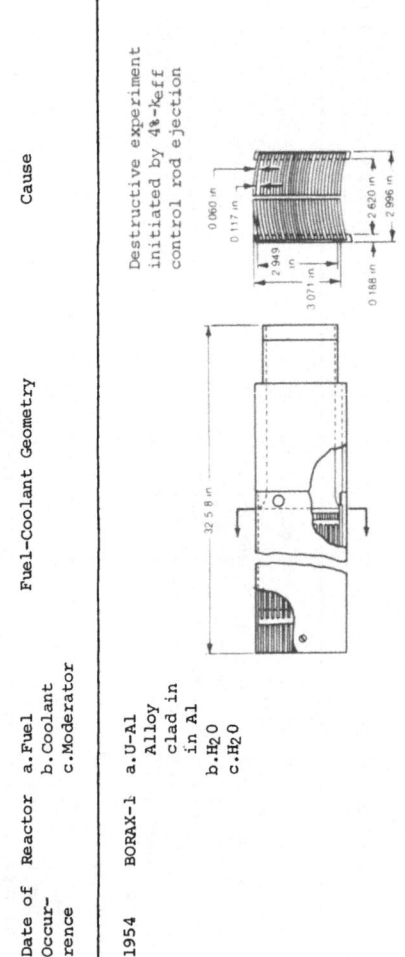

INEL-A-8302

Description of Events	Primary Factors Influencing Reactor System Damage
Analysis of the accident indicated that most of the fuel plates melted or partially vaporized. However, postaccident debris indicated that although the U-Al fuel melted during the transient, the cladding remained solid. The total energy release was 135 MW-s, resulting in a pressure peak of about 650 atms. The reactor tank was burst by the blast and most of the shield tank contents were ejected into the air. The control rod mechanism (mounted on a heavy plate), weighing approximately one ton, was thrown 30 feet into the air. The shutdown mechanism was due to core mechanical expansion and coolant boiling.	a. Although high-speed motion pictures indicated a light flash lasting 0.003 s as peak power was reached, the flash was gone before any material was ejected from the reactor tank; thus, indicating that the mechanical damage was a result of a thermally-induced steam explosion rather than a chemical explosion. b. Postaccident analysis indicated that the fuel was molten at the time of interaction, while the Al cladding remained in the solid state. c. Extensive fuel fragmentation occurred.

TABLE I

Date of Occurrence	Reactor	a. Fuel b. Coolant c. Moderator	Fuel-Coolant Geometry	Cause
1/3/61	SL-1	a. 93% enriched U^{235}, clad with Al b. H_2O c. H_2O	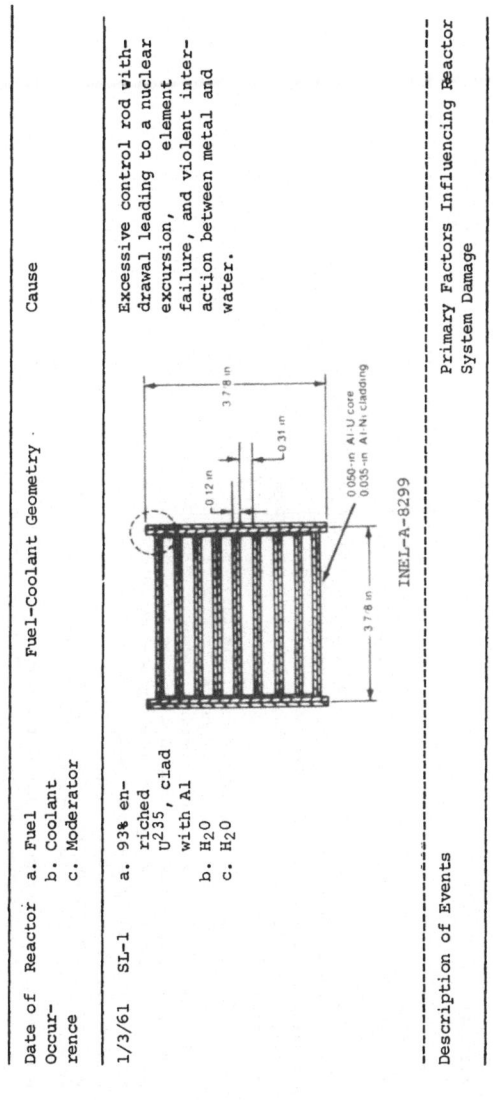	Excessive control rod withdrawal leading to a nuclear excursion, element failure, and violent interaction between metal and water.

INEL-A-8299

Description of Events	Primary Factors Influencing Reactor System Damage
Analysis of the accident indicated that the excess reactivity resulted in partial fuel element meltdown and vaporization. Intermixing of the vapor driven solid/molten metal with water led to violent steam formation and some metal-water chemical reaction. The formation of a steam void terminated the nuclear transient. However, the steam pressure (estimated to have reached 700 atm) caused the pressure vessel to rise approximately 9 feet above its normal position, shearing off all piping in the process, and ejecting the loose control rod plugs upward, resulting in their penetration in the ceiling above. Two of the fatalities were instantaneous, as a result of the mechanical blast damage. The third fatality was due to radiation exposure and flying debris. The mechanical damage was the result of a shock-type metal-water thermal explosion, being approximately 1% of the total nuclear energy released.	a. Essentially all mechanical damage to the reactor system can be attributed to the thermally-induced steam explosion. Calculations indicated that at the time of the steam vapor expansion, the Al cladding was solid, while the U metal was partially vaporized.

TABLE 1

Date of Occurrence	Reactor		Fuel-Coolant Geometry	Cause
	a. Fuel b. Coolant c. Moderator			
11/5/62	SPERT-10	a. U-Al Alloy b. clad in Al b. H2O c. H2O	 INEL-A-8301	Destructive experiment initiated by a power excursion.

Description of Events	Primary Factors Influencing Reactor System Damage

The destructive test was instrumented with a special capsule to measure transient power, energy, fuel-plate surface temperature, moderator flow, pressure, and strain. The data obtained indicate a period of 3.2 ms and a peak power of 2300 MW(t). Pressures of (270 atm) were recorded. Approximately 35% of the core experienced melting with all 270 fuel plates in the core having undergone some degree of melting.

Postaccident analysis indicated that approximately 20 kg of a sponge-like metallic material fragmented to sizes from 0.003 inch to a few inches in diameter. Chemical analysis of this material indicated a content, by weight, of 65% Al, and 6.7% U; the rest being insoluble residue (Fe, etc.). The extent of Al-H$_2$O chemical reaction was considered small with respect to the pressure pulse. The blast pressure was considered to be caused by a steam explosion resulting from thermal interaction between the molten fuel plates and the water.

The maximum surface temperature at the time of the destructive pressure pulse was estimated to be about 600°C; the interior portion of the U-Al fuel was about 1200°.

a. The metal/water chemical reaction did not appear to be the primary cause of the observed pressure pulse.

b. Fuel-plate analysis indicated that approximately 35% of the core was molten at the time of the blast.

c. A steam explosion resulted, considered to be caused by a thermal interaction between metal and water.

An overview of current international MFCI research
activities is given in Reference *23* and illustrated in
Table II for light water reactors (LWR's), and in Reference
24 and Table III for liquid metal fast breeder reactors
(LMFBR's). However, the safety concerns for the two systems
are somewhat different. With respect to LMFBR's, the questions of vapor explosion potential and milder MFCI leading
to coolant voiding (with a potentially positive void coefficient) both are of primary importance. For LWR's, however, the effects of coolant expulsion are not as critical
in comparison, due to an overall negative void coefficient.
Thus, for LMFBR's, two important concerns exist; vapor explosion potential and milder MFCI-induced voiding; whereas,
for LWR's, violent vapor explosions are of primary interest.
As stated in the introduction, this report is limited to a
discussion of various modeling concepts associated with
MFCI-induced vapor explosions rather than a general discussion of all aspects of molten fuel-coolant interaction.*
In the following section, a critical review is presented of
two overall vapor explosion model concepts in view of the
basic physical phenomena assumed in each model.

III. VAPOR EXPLOSION MODELS

Basically, two different approaches have been adopted
in assessing the vapor explosion problem; the equilibrium
thermodynamic models that estimate the masimum work potential available, and the transient heat transfer and fluid
dynamics models that attempt to describe the actual phenomena
and rate processes involved.

The thermodynamic approach was first employed by Hicks
and Menzies(*27*) to estimate an upper limit on the expansion
work of the coolant due to an MFCI. This work is calculated
to be equal to the change in internal energy of the fuel during an isentropic expansion from a compressed state, defined
by the amount of energy added to the core fuel and to an expanded state, defined by the boiling temperature of the fuel

---

*   Present discussion is limited to the highly improbable
        core-meltdown accident with accompanying potential for
        a thermal vapor explosion, and should not be considered
        applicable to other hypothesized loss-of-coolant acci-
        dents(*25,26*).

TABLE II

STEAM EXPLOSION RESEARCH PROGRAMS WITH RESPECT TO CORE MELTDOWN ASSESSMENT PROGRAMS (23)

Phase	Sandia Laboratories	Joint Research Center, Ispra, Italy		Argonne National Laboratory	
Test configuration	Stationary vapor-blanketed droplet in closed chamber	Shock tube	Drop test in 300-closed tank	Drop tests in closed vessel	Various modes of forced contact
Hot phase	Steel, corium	UO_2, steel, zircaloy-4, corium	UO_2, steel, zircaloy-4, corium	Mineral oil	Aluminum
Cold phase	Water	Water	Water	Freon-22	Water
Temperature hot phase (°C)	T_{mp} to $(T_{mp} + 500)$	1500 to 2900	1500 to 2900	100 to 220	300
Mass of hot phase (kg)	0.02	0.12 to 0.15	3	5	0.10
Mass of cold phase (kg)	2	0 to 22	0 to 200	10	1
Temperature of cold phase (°C)	25 to 90	20 to 220	20 to 220	- 40	25
System pressure (atm)	1 to 5	1 to 25	1 to 25	1 to 50	1

TABLE III

EXPERIMENTAL LMFBR FUEL-COOLANT INTERACTION RESEARCH PROGRAM (24)

Laboratory	UO	Na	Contact Mode	Test Conditions
			Out-of-Pile Experiments	
Grenoble France	5 3 to 5 kg	4 225 g	UO_2 into Na Na into UO_2	Rod failure, joule heated Pool geometry, induction heated
Japan	20 g	30 to 200 g	UO_2 into Na	Dropping, radiant heated
CNEN, Italy	10 g	7 g	UO_2 into Na	Rod failure, joule heated
Ispra, Italy	3,000 g	70 g	UO_2 into Na	Dropping, radiant heated
ANL, U. S.	7 to 10 g 30 g 500 g	300 g 5 to 10 g 5 to 10 g	UO_2 into Na Na into UO UO_2 into Na	Dropping, inducted heated Dropping, induction heated Dropping, thermite heated

TABLE III

In-Pile Experiments

Laboratory	UO	Test Conditions
Petten, Netherlands	100 g	Single-rod, fresh fuel, loss-of-flow (LOF) simulation
ANL (Series):		
(S)	28 to 286 g	1 to 7 rods, fresh fuel, transient-overpower (TOP)
(H)	68 to 472 g	1 to 7 rods, fresh and irradiated fuel, TOP (50¢/s)
(E)	45 to 472 g	1 to 7 rods, fresh and irradiated fuel, TOP ($3/s)
(L)	456 to 470 g	7 rods, fresh and irradiated fuel, LOF
(R)	200 to 1200 g	1 to 7 rods, fresh fuel, LOF
Sandia	64 g	1 rod, fresh fuel, with and without Na, prompt burst

at atmospheric pressure. By use of a different equation-of-
state for sodium coolant, Judd(28) obtained a somewhat higher
($\simeq$ 30%) estimate of the ultimate work potential. However,
such thermodynamic approaches do not describe the necessary
conditions for explosive vaporization, nor the details of
the physical phenomena or rate processes involved. A more
mechanistic approach to the problem has been the development
of several transient models, the two most important being
the spontaneous nucleation and pressure-induced detonation
models. A critical review of these two transient model concepts is presented in the following sections.

A.   Spontaneous Nucleation

As illustrated in Figure 3, the spontaneous nucleation
model proposed by Fauske(29,30) considers that for a large-
scale vapor explosion to occur, the conditions for liquid-
liquid contact must exist such that the contact interface
temperature established between the molten fuel and coolant
must exceed that for spontaneous nucleation (Figure 3 (a)).
The contact interface temperature (T_I) in the absence of
solidification of the hot material can be described as

$$T_I = \frac{T_H \, (k/\sqrt{\alpha})_H + T_c \, (k/\sqrt{\alpha})_c}{(k/\sqrt{\alpha})_H + (k/\sqrt{\alpha})_c} \tag{1}$$

where

$\qquad$ T = temperature

$\qquad$ k = thermal conductivity

$\qquad$ α = thermal diffusion

$\qquad$ H = hot material

$\qquad$ c = coolant

To assess the temperature at which spontaneous nucleation occurs (T_{SN}), Volmer's well-known rate equation is
used, such that

$$J = const \, exp \, (-W/KT) \tag{2}$$

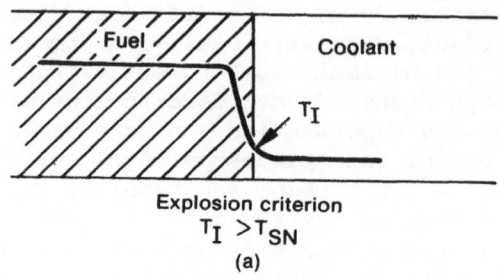

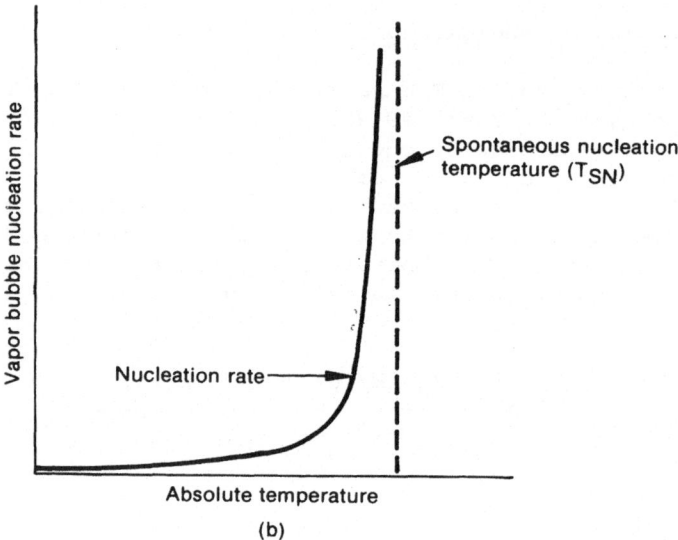

Figure 3.   Illustration of Spontaneous Nucleation Model

where

J = the rate of bubble nucleation per unit volume

K = Boltzmann's constant

W = reversible work of formation of a critical
vapor embryo in the liquid and is given by

$$W = 16\pi\sigma^3/3(P_v - P_\ell)^2 \tag{3}$$

where

σ = surface tension of the pure liquid

p_v = vapor pressure

p_ℓ = liquid pressure

As shown in Figure 3(b), the nucleation rate remains small until the temperature of the liquid reaches a critical value, wherein J increases rapidly. For vapor nucleation in the bulk of a pure liquid, this temperature limit is referred to as the homogeneous nucleation temperature (T_{HN}) *
and can be approximated as about 90% of the thermodynamic critical temperature, T_c ($\approx$ 647 K for water and $\simeq$ 2570 K for sodium). However, for the case of partial liquid contact with another substance, the effect of the wetting characteristics at the contact interface also must be considered, such that Equation (3) is multiplied by a wetting factor, expressed as

$$f(\theta) = \frac{2 + 3 \cos \theta - \cos^3 \theta}{4} \qquad (4)$$

where θ= the contact angle established between the two materials. For perfect wetting, θ = 0° and $f(\theta)$ = 1, which gives the upper limit of the spontaneous nucleation temperature (T_{SN}); that is, the homogeneous nucleation temperature. Only for a high degree of nonwetting is T_{SN} significantly lower than T_{HN}, where, for example, as θ approaches 180°, T_{SN} approaches the saturation temperature.

As can be seen by inspection of Equation (1), the interface temperature approaches that of the hot material when the thermal conductivity of the coolant is relatively low, compared to that of the hot material. As a result, forced contact between such molten fuel and coolant may lead to the condition that $T_I > T_{SN}$, which is the situation for molten

---

* The homogeneous nucleation temperatures calculated from Equation (2) for various fluids are: Freon-22 (326 K), N-pentane (421 K), ethonol (468 K), sodium (2300 K), and water (575 K).

UO_2 quenched in water. If sufficient heat transfer surface
area is available through the process of fuel fragmentation
and intermixing, and if the system is constrained such that
a rapid rate of pressure buildup cannot be relieved, then a
shock-type vapor explosion is postulated to occur. However,
it has been argued(29) that fission gases, radiation, and
other nucleation aids in a reactor environment tend to initiate boiling prior to the spontaneous nucleation threshold
being reached, such that relatively mild pressurization
events would normally occur. However, energetic vapor explosions of the type observed in the SPERT-ID, BORAX-I, and
SL-1 events can occur if intimate contact for a constrained
system is established. For LMFBR materials such as UO_2
fuel and Na coolant, such explosions are ruled out 'a priori'
for the reactor conditions of large UO_2-Na masses, since in
this case, the contact temperature is calculated to be well
below that for spontaneous nucleation of sodium. Only for
the laboratory case, where a small mass of Na might be en-
trapped by UO_2 and slowly heated to its spontaneous nucleation temperature, is a vapor explosion considered possible.
Armstrong's experiments(31,32) have been interpreted in Reference 29 in this manner. However, in a recent paper by
Anderson and Armstrong(33), both UO_2 into Na and Na into UO_2
are considered to behave in a similar manner; both producing
laboratory-scale vapor explosions under favorable conditions
(to be discussed later). Thus, the statement that vapor explosions have not been observed for the UO_2 into Na experiments appears to be subject to interpretation.

Other experimental findings that tend to discount the
spontaneous nucleation-contact interface criterion are the
R-22[*] water experiments of both Anderson and Armstrong(33)
and those of Enger and Hartmann(10). With respect to the
latter experiments, the contact temperature was pointed out
to be below that for homogeneous nucleation (T_{HN} (R-22) =
53°C), yet vapor explosions occurred. Fauske(29) attributed
this discrepancy to transient variations in the interfacial
surface energy (that is, dynamic surface tension effects)
during the initial stage of contact. However, since R-22
has been shown to spread easily on cold water(34), the spontaneous nucleation temperature would be expected to be near
the homogeneous nucleation temperature.

---

[*]   Called Refrigerant-22 or Freon-22, the chemical formula
       being $CHClF_2$.

   While the question still remains whether or not achievement of the spontaneous nucleation temperature is a necessary
condition for the occurrence of vapor explosions, several
interpretations have been postulated to account for an exper-
imentally-observed delay period from time of contact to onset of rapid vapor production and pressurization. Henry
and Fauske(35-37) have attempted to account for such a delay period in the context of the time required to attain
spontaneous nucleation from the onset of liquid-liquid contact. Although the time scale for establishment of the interface temperature is 10^{-12}s, assessed from the relaxation time
for thermal vibration propagation (α/c^2, where c is the velocity of sound in the conducting material), Henry et al(35-
37) hypothesized that vapor bubble nucleation cannot proceed
until a thermal layer has developed in the coolant phase
which is sufficiently thick to support a vapor embryo of the
critical size (that is, $R_c = 2 \sigma /\Delta P$). The time scale for
commencement of bubble nucleation after contact is assessed
considering the pressurization effects of liquid-phase expansion during development of such a thermal layer. The
problem in such an assessment is one of determining the relief time and associated thermal layer thickness for the
formation of critical size embryos (here called the critical
thermal layer). For the case where the critical thermal
layer is developed before pressure relief (for example, a
fluid with a high diffusivity and a low velocity of sound,
and a system with a relatively long distance to the free
surface), a vapor explosion is considered to be temporarily
suppressed if the pressurization exceeds the critical thermodynamic pressure, P_c.

   For the opposite situation, in which rapid pressure
relief occurs, the time period (t_n) elapsed between contact
and the inception of a critical size bubble embryo is con-
sidered(35-37) to be governed by the sum of three waiting
times

$$t_n = t_a + \frac{1}{JV} + \frac{2\ell}{c}$$
(5)

where t_a is the acoustic relief time for liquid phase thermal
expansion (which occurs during the initial period of contact
between the hot and cold fluids), $\frac{1}{JV}$ is the waiting time until
bubble nucleation, and $2\ell/c$ is the time period necessary for
a pressure gradient to develop in the vicinity of the nucle-

ated bubbles so that inertial bubble growth is possible.   In
Equation (5), J is the nucleation frequency per unit volume,
V is the bubble volume, ℓ is a characteristic distance to a
free interface, and c is the velocity of sound.

The time for the vapor bubble to grow (t_g) from nucleation to the limit of its stable region can be calculated
from the inertial bubble growth equation such that

$$t_g = \frac{R_b(t) - R_c}{\sqrt{2/3} \ (\Delta P/\rho_\ell)} \tag{6}$$

where

$R_b(t)$ = time-dependent bubble radius

R_c = critical bubble radius

ρ_ℓ = liquid density

ΔP = average driving pressure from bubble ·
inception to the maximum stable bubble
limit

During this growth period, additional bubbles can form
such that if the contact temperature is considerably higher
than the spontaneous nucleation value, the nucleation rate
is rapid enough to cause bubble interference.   Such a condition will lead to vapor blanketing between the two previously contacting liquids which, in turn, reduces the energy-
transfer process.

To assess whether such vapor nucleation will lead to
either explosive boiling or simple vapor blanketing of the
hot surface, a maximum site density (based on pressurization
sufficient to suppress further nucleation) is compared with
an interference site density ($N = D_B^3$).   By evaluation of
the maximum stable bubble diameter (D_B) from the ideas outlined above, Henry et al(35-37) found that larger-diameter
cold droplets submerged in a hot liquid result in site density interference leading to protective vapor blanketing,
rather than in explosive boiling.   An analytic prediction of
this behavior is shown in Figure 4 for single droplets of
Freon-12 (cold phase) submerged in oil (hot phase) and com-

pared with experimental data. With respect to large cold
droplet systems, it is hypothesized that film boiling will
occur at high interface temperatures because of a well-
developed thermal boundary layer and the large frequency of
nucleation. As a result, such large cold droplets should
remain in a film boiling condition until either they break
up into smaller droplets or film collapse of the surrounding hot fluid occurs. If, at that time, the interface temperature still exceeds that for spontaneous nucleation, a
potentially explosive interaction is considered possible.

Several authors have critiqued the work of Henry et al
(35-37) and proposed alternate arguments for a period of
stable film boiling for the condition of $T_I > T_{SN}$. Board
and Hall(38) have proposed that at temperatures above T_{SN},
rapid bubble growth and coalescense quickly leads to the
formation of a stable vapor film that limits the vapor generation process (which agrees with Henry's arguments). However, they propose that at temperatures just below T_{SN},
bubble growth and coalescence conditions are such that the
total volume of vapor generated is greater than that above
T_{SN}; thus, the condition where $T_I < T_{SN}$ should lead to a
more energetic situation than when $T_I > T_{SN}$. Board and
Hall(38) consider the Leidenfrost point (T_{Leid})*, or minimum
temperature for stable film boiling (assumed to be below T_{SN}),
as a better indicator of a threshold temperature for an
energetic interaction. Besides questions relating to the
validity of the interface-spontaneous nucleation criterion,
arguments also have been present that are at variance with
other aspects of the model and with the interpretation of
associated simulant fluid experiments.

---

* From a thermodynamic viewpoint, one estimate of the
  minimum wall temperature to sustain film boiling (that
  is the Leidenfrost temperature) is 27/32 T_c (39). However, for a pool boiling situation, the minimum temperature for film boiling may be somewhat higher for small
  spheres(40) and large subcooling(41), based on hydrodynamic theory. At this time, an 'a priori' assessment
  of T_{Leid} cannot be made with confidence, although for
  well-wetted systems, the thermodynamic approach appears
  to correlate data quite well(42). For poorly-wetted
  systems, T_{Leid} may approach the saturation temperature
  T_{sat} (42,43).

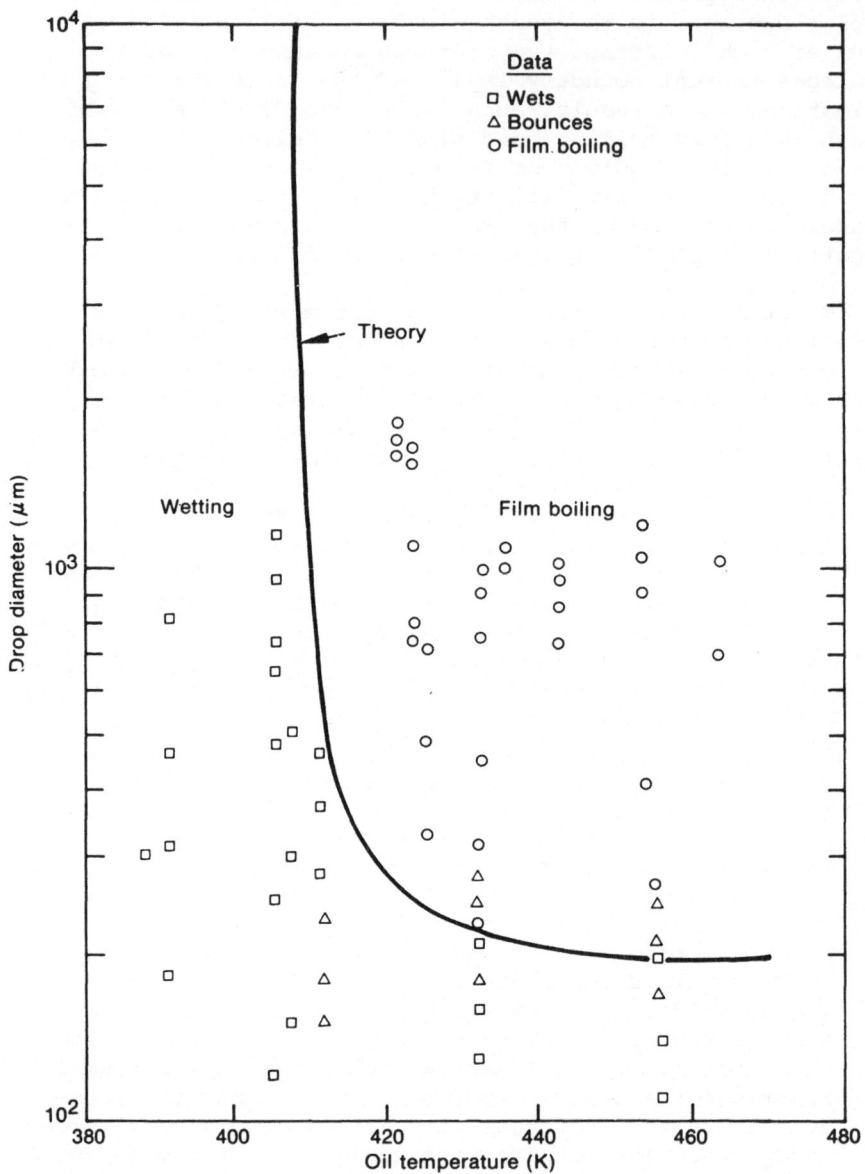

Figure 4.   Small Drop Interaction Behavior for Film Boiling
            and Capture of Freon-12 on a Mineral Oil Surface (37)

W. B. Hall(*44*) has presented calculations for bubble growth with acoustic loading, indicating that it is not necessary to await relief in order for bubble growth to proceed, the time period being dependent on the site density (N). Such results are in basic disagreement with the arguments of Henry et al(*35-37*) for suppression of nucleation due to acoustic loading. However, Hall's formulation appears to be correct since an acoustic loading term is coupled directly with the Rayleigh bubble growth equation. The acoustic loading term is given as

$$\Delta P = 4\pi N \ R^2 \ \dot{R} \ \rho C \tag{7}$$

where

N = site density

R = radius

$\dot{R}$ = velocity

ρ = density

C = speed of sound

Another consequence of the spontaneous nucleation model is that if the system pressure (P_{sy}) is above the thermodynamic critical value (P_c), discrete phase change will not occur; thus, an explosion is considered to be prevented if $P_{sy} > P_c$. However, as illustrated in Figure 5, work is a path function such that rapid heating of a supercritical fluid prior to inertial relief of the constraining system can lead to an expansion process into the vapor regime that can be explosive, the rate processes being the determining factor.

Other than Henry's Freon into oil (or water) studies (where the minimum Freon droplet size is assessed for explosive vaporization), no detailed coupled fragmentation-spontaneous nucleation model has been developed to date that answers questions concerning the mechanism and kinetics of the fragmentation-intermixing-heat transfer processes (for a large-scale system) versus the kinetics and energy associated with such nucleation. A complete description of the processes involved from nucleation fragmentation, intermixing,

and eventual explosion appears to be necessary at this time.
Cho et al(45,46) have begun to investigate such questions,
particularly mixing requirements that are discussed in a
later section.

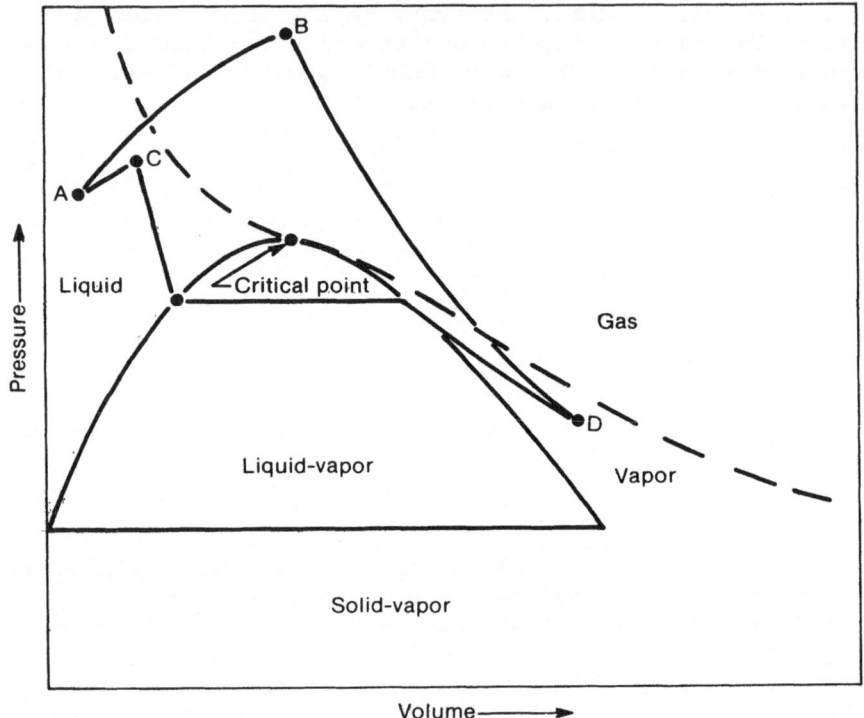

Figure 5.   Illustration of Various Expansion Processes for
            a Rapidly-heated Supercritical Fluid (Paths A-B
            and A-C are prior to Inertial Relief), Indicating
            Work is a Path Function.

     Several other important factors have yet to be incor-
porated into this model concept with respect to assessing
spontaneous nucleation for actual reactor materials and system conditions.   For example, the work potential associated
with a vapor explosion is mass-dependent, yet the amount of
interacting material has not been discussed in the context

of this model. With respect to the kinetics of phase change,
it has been assumed that for perfect contact between two
interacting fluids, Volmer's well-known rate equation applies;
thus, assessing homogeneous vapor nucleation for the liquid
coolant but ignoring the effects of homogeneous solid nucleation in liquid fuel. In References 47 and 48 it has been
demonstrated that Volmer's equation also describes the kinetics of solid phase nucleation in the melt and that under
perfect contact quenching conditions, liquid UO_2 will undergo similar homogeneous solid crystal nucleation(49,50,51).
This important effect has not been accounted for in the model
or the simulant fluid experiments of Henry et al(35). It
is felt that the effects of simultaneous solid-phase and
vapor-phase nucleation for two dissimilar fluids has yet to
be dealt with in a vigorous manner and that the present understanding of vapor nucleation from solid surfaces indicates
that nucleation kinetics for such a simultaneous process may
be quite different than for liquid-liquid systems.

Besides the concept that spontaneous vapor bubble nucleation is a necessary condition to cause explosive vaporization,
Board and Hall(38,52,53) have suggested an entirely different approach for explosive vaporization. Their model concept and variations of it are discussed in the following
section.

B.   Pressure-induced Detonation Model

Board and Hall(38,52,53) have proposed the theory of
explosive fuel-coolant interaction by pressure-induced detonation, similar to that of detonating chemical explosions.
As illustrated in Figure 6, the model assumes that a strong
shock front propagates steadily through a region of coarsely-
mixed molten fuel and coolant, the initial pressure trigger
being considered sufficient to cause collapse of any pre-
existing vapor. As the pressure wave passes the interaction
region, the flow velocity differential between the dense fuel
and lighter coolant are considered sufficient to cause fine-
scale fragmentation. As a result, the front leaves behind
finely fragmented fuel in intimate contact with the coolant,
eventually resulting in vaporization sufficiently rapid to
cause shock pressurization and explosion. Some uncertainties with respect to the model are: the source of the initial
pressure pulse, the requisite condition of a predispersed
coarse mixture, and the mechanism for rapid, fine fragmentation by dynamic instability.

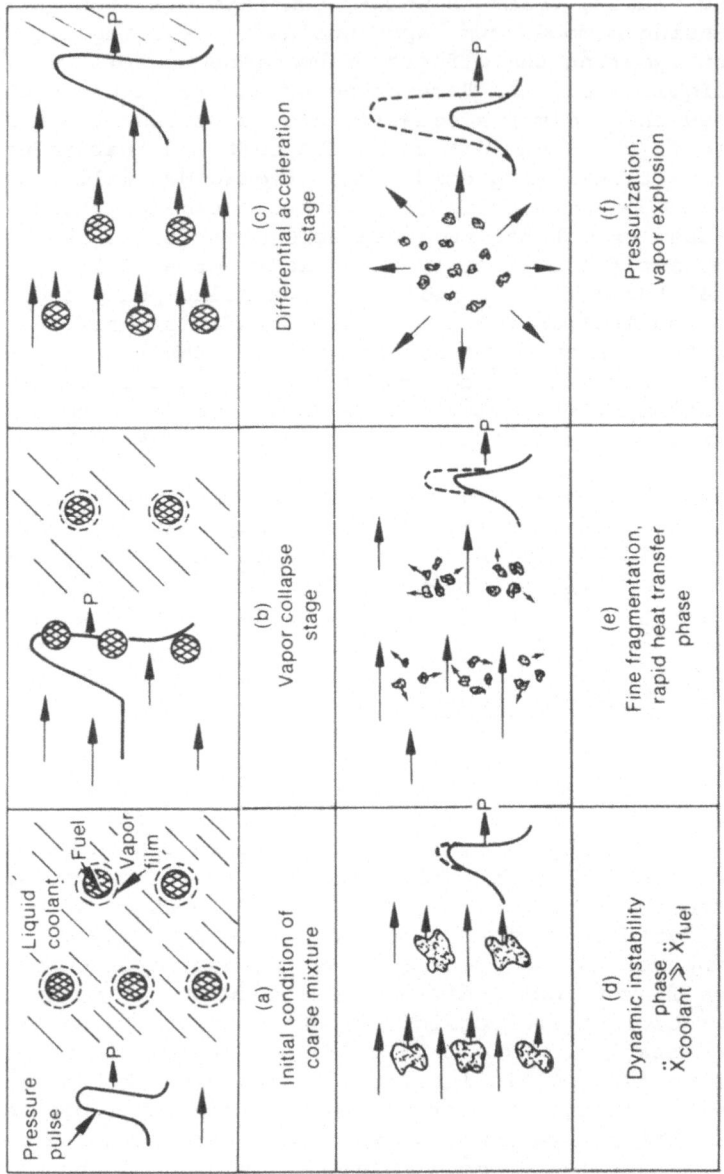

Figure 6. Descriptive Illustration of Pressure Detonation Model

At present, the requirement of rapid, fine fragmentation and intermixing is receiving considerable attention, since without such rapid fragmentation and intermixing, increased pressurization and eventual detonation ceases. The time for breakup is important to the analysis, since slow fragmentation (after the pressure pulse has traveled downstream of the interaction zone) would not result in superposition of pressurization, but rather, a series of pressure waves that would eventually damp out.

To calculate the fragmentation time, the semi-empirical relation of Simpkins and Bales(54) was used for the breakup of a single droplet accelerated in a gas velocity field. The resulting breakup time is calculated to be

$$t_b = 44 \ Bo_{L-V}^{-1/4} \ \sqrt{\frac{\rho_d}{\rho_f}\left(\frac{R_d}{V_r}\right)} \tag{8}$$

where

ρ_d = drop density

ρ_f = fluid density

R_d = drop radius

V_r = velocity of the drop relative to the surrounding medium

Bo_{L-V} = bond number for a liquid-vapor system that is calculated as

$$Bo_{L-V} = \frac{\rho_d g \ R_d^2}{\sigma} \tag{9}$$

where

σ = interfacial tension

g = acceleration of the drop due to drag

To estimate the relative velocity after the shock front passes the drop, the mixture velocity at the Chapman-Jouguet point is used. For gas shocks, this is satisfied only if

the drop is accelerated by the drag of the surrounding fluid
and the fluid velocity remains constant through the
fragmentation zone. However, these assumptions may not
apply to the liquid-liquid system. To account for the
situation of a liquid-liquid mixture, Bankoff et al(55,56)
suggested that a modified Bond number (Bo') be used, such
that

$$Bo'_{L-L} = Bo_{L-V} \left[\frac{(\rho_d - \rho_f)\sigma_d}{(\rho_d - \rho_f)\sigma_d} \right] \tag{10}$$

where σ is the surface tension of the droplet (d) and fluid
(f). Such a correction is relatively minor in importance.
However, Bankoff et al(55,56) also investigated the effects
of mass ratio on the relative velocity obtained between the
droplet and coolant fluids and the time of breakup. Basically, a momentum balance is written for the coolant and
droplet fluids to determine the time-dependence of the relative velocity (V_r). The resultant expression for this
velocity is

$$\frac{1}{V_r^2} \frac{dV_r}{dt} = - \frac{C_D A \rho_f}{2} \left[\frac{1}{m_d} + \frac{1}{m_f} \right] \tag{11}$$

where

C_D = drag coefficient (assumed to be 2)

A = projected area of the drop

m = mass

Considering both the mass effect for a multiple droplet configuration and the Bond number for liquid-liquid systems,
the time for breakup is calculated as

$$t_b = 33 \; Bo'_{L-L}^{-1/4} \sqrt{\frac{\rho_d}{\rho_f}} \left(\frac{R_d}{V_r} \right) \left(1 + \frac{m_d}{m_f} \right) \tag{12}$$

As can be seen, the mass ratio effect predominates over that
introduced by the modified Bond number for liquid-liquid
systems.

A comparison of the time for breakup using the Board-Hall versus the Bankoff assumptions indicated that the breakup time estimated by Board and Hall (single droplet) can lead to fast fragmentation which is necessary for sustaining a detonation wave, whereas if the mass ratio (multiple droplets) effects are considered, the time for breakup is much slower and does not meet the criteria for sustaining the pressure wave. However, it is important to note that the semi-empirical relation of Simpkins and Bales (54) used in both calculations was developed, based on data for liquid drop breakup in a gas environment; thus, Bankoff's arguments may not apply to the liquid-liquid system.

Contrary to such arguments that hydrodynamic fragmentation in a liquid-liquid system is more difficult to achieve than for the gas-liquid case, Patel and Theofanous(57) have shown experimentally that hydrodynamic breakup of mercury in water under shock tube conditions leads to rather efficient fragmentation, which appears more likely to occur than the expected results based on the theory for gas-liquid systems. For liquid-liquid systems, Taylor-type* instabilities are considered to be the controlling factor.

Further investigation with respect to liquid-liquid systems appears to be warranted since without a quantitative knowledge of such fragmentation, the detonation concept cannot be accurately assessed. It is important to note that thermal effects should be accounted for in any final assessment of the fragmentation process since, in reality, a highly non-isothermal situation exists.

Williams(58) also investigated the assumptions of the Board-Hall model and concluded that the probability of fine-scale fragmentation is greatly overestimated. His critique was made based on an assessment of the magnitude of the initiating pressure shock necessary to accelerate the fuel to a velocity corresponding to the Chapman-Jouguet point.

─────────────────────

* A Taylor instability is that which is due to an acceleration perpendicular to the interface of two fluids of different densities, where relative velocity effects between the fluids are unimportant.

For the UO_2-Na fluid system, an initial pressure pulse of
80 to 700 bars was assessed, corresponding to a vapor void
fraction of 0.5 to 0.1, respectively. The magnitude of
such a pulse implies that the initiating or trigger event
would be difficult to obtain and hardly less destructive
than the vapor explosion itself. For heavy metal-water
systems, similar results can be expected, since such material properties as the speed of sound and isentropic bulk
modulus are not that dissimilar from the UO_2-Na system.
Williams also discussed the dispersive characteristics of
shock waves in the context of the Board-Hall model. He
points out that sharply-defined pressure waves undergo multiple partial reflections at the interface between mixture
constituents, resulting in an attenuated wave. Multidimen-
sional effects also must be considered with respect to reactor systems. Investigations(58,59) of the Board-Hall
assumptions also indicate that the reaction zone must be
rather large, on the order of reactor system dimensions;
thus, the fact that most small, out-of-pile experiments
do not result in shock-type vapor explosions may not be
relevant to the question of whether such explosions can
occur in a nuclear reactor. Recent calculational results
of Sharon and Bankoff(59) indicate that for high-pressure
shocks ($\sim$ 3,000 to 8,000 atm), a rather long relaxation
length (several meters)* is required for good intermixing
of fuel and coolant (the relaxation length being defined
as the distance from the leading edge of the shock front to
the point of velocity equilibrium between fuel and coolant).
With respect to geometry considerations, it thus appears
that a preliminary critique of the detonation concept indicates rather large system dimensions would be required (for
reactor fuel-coolant materials) to meet detonation conditions
(that is, dimensions greater than those common to reactor
design concepts). Prototypic experiments ultimately may be
necessary, although the cost would be phenomenal; therefore,
continued analytical investigation at this time appears to
be appropriate.

Gunnerson and Cronenberg(41) have investigated the
initial blanket-coarse mixture geometry assumed by Board-
Hall, wherein a vapor blanket is considered initially to
surround the fuel particles. The results of their work are

---

* Dependent upon the drag coefficient, particle (drop)
size, densities, and breakup time.

shown in Figure 7 and indicate that particle diameter has a large effect on the minimum temperature necessary to sustain film boiling. This minimum temperature is predicted to increase with decreasing particle size and is estimated to be quite large for particle sizes of interest in MFCI analysis. As a result, the initial vapor blanketing condition assumed by Board and Hall may be unrealistic, depending upon the particle size characterization assumed for the initial coarse mixture. Also, it should be noted that many dropping experiments of molten metals into water(2,7,60-62) have resulted in extensive fragmentation in the absence of any significant velocity differential. As discussed in the following section, other heat-transfer-governed mechanisms (for example, boiling and solidification) may account for a rapid, fine fragmentation-intermixing process, with attendant rapid heat transfer sufficient in nature to induce acoustic pressurization of the coolant and eventual downstream explosion. If it is demonstrated that this is true, the rather severe initial pressure trigger assumed by Board may not be necessary. It is suggested here that the kinetics of such alternate fragmentation mechanisms be investigated in the context of this model.

Recently, Anderson and Armstrong(33) have proposed a model similar to that of Board and Hall(52). They consider that a vapor explosion occurs as a result of three independent steps: (a), an initial mixing phase; (b), a trigger and growth phase; and (c), a mature phase in which a shock wave accelerates the two liquids into a collapsing vapor layer, causing high-velocity impact that finely fragments and intermixes the two liquids.

Of particular interest are the experimental results of Anderson and Armstrong(33) which led to this concept. In one series of experiments, unrestrained drops of a hydrocarbon refrigerant (R-22) were poured into a water-filled container. The results of these tests showed that:

1.   pressurization rise times are relatively slow (0.5 to 4 ms)

2.   The measured pressure history is strongly influenced by the geometry of both the water-filled tank and the Freon-22 mass

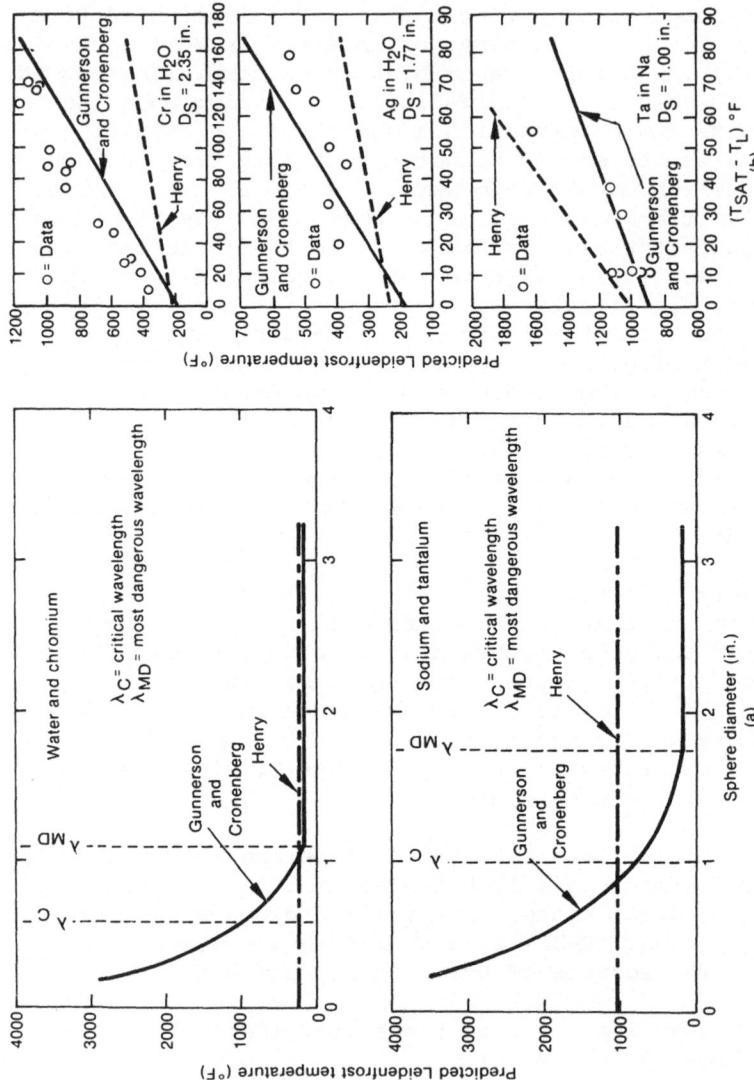

Figure 7. Predicted Leidenfrost Temperature Versus (a), Sphere Diameter; and (b), Degree of Subcooling for Three Metal-coolant Systems (41)

3.   Interaction behavior changes at a
      water-bath temperature of about 75°C.

4.   peak-generated pressures at a given
      delay time were doubled when a small
      quantity of a dye was added to the
      Freon-22.

5.   Delayed, high-pressure events occurred
      in all water tests with water temper-
      atures between 75°C and the maximum
      attainable system temperature of 99°C.

Such results indicated two types of interactions due
to surface boiling characteristics. For the high-temperature
experiments, a period of film boiling enabled the R-22 mass
to penetrate deeply into the water before a reaction was
initiated, such that the inertial constraint of the water
depth contributed to the generation of high pressure (rather
than attributing this to the criterion of $T_I > T_{SN}$). Thus,
Anderson and Armstrong(33) postulated that high-pressure explosions also might be produced with somewhat lower temperatures if R-22 were suddenly released well below the water
surface. To test this hypothesis, R-22 was fully submerged
in water by first enclosing it in a rubber prophylactic.
The results of such constrained tests can be summerized as
follows:

1.   Subsurface release of Freon-22 into water
      below 60°C produced a qualitatively different
      kind of reaction than release into water
      above 70°C.

2.   Low-temperature reactions were characterized
      by the immediate interaction at the Freon-
      water interface.

3.   High-temperature reactions proceeded in two
      steps: first, the Freon-22 surface went
      into a film-boiling condition; second, film
      destabilization by a surface mixing wave
      resulted in explosive reaction.

4.   The peak pressures in each test were directly
      proportional to the delay time between the
      vapor film formation and the appearance of the
      mixing wave.

5.   The relative efficiency ranged from 5%
     in low-temperature experiments to 60%
     in high-temperature experiments.

Although the conditions proposed for explosive interaction are consistent with such experimental findings, so,
too is the spontaneous nucleation concept, in that the
high-temperature tests were the most violent and satisfy
the condition that $T_I > T_{SN}$. However, a thermodynamic prediction of the destabilization temperature for film boiling
can be of similar magnitude as the threshold temperature
for spontaneous nucleation(42); thus, such experiments can
be interpreted in light of both factors. The fact that a
wave-type disturbance appeared to trigger events lends added
support to the shock-induced film boiling destabilization
concept. Although Anderson and Armstrong(33) present calculations supporting fine-scale intermixing of the two fluids,
questions arise as to the assessed mixing energy requirements and the kinetic energy available upon vapor film collapse. This uncertainty arises as a consequence of assessing impact velocities and extent of fluid intermixing at the
time of interaction.

Besides pressure or shock-wave-induced film boiling
destabilization as a trigger for violent fuel-coolant interactions, Gunnerson and Cronenberg(40,42) have indicated that
such destabilization also can be induced by thermal conditions where an assessment of the minimum temperature for
film boiling is assessed from thermodynamic considerations.
The results of their work indicate that a strong shock wave
may not be a necessary condition to cause initial vapor collapse, although such a shock wave may be necessary for fine-
scale fragmentation and intermixing requirements.

From the preceding discussion, it is apparent that a
number of questions remain relating to the previously
discussed model concepts for vapor explosion. Such questions
are summarized in the next section.

IV. SUMMARY OF UNRESOLVED QUESTIONS

Some important questions that have not been resolved to date with respect to interpretation of experimental and modeling results are:

1. Did the experiments of Armstrong et al(*31,34*) with UO_2 into Na produce explosions similar to the Na into UO_2 tests?

2. Are dynamic surface tension effects for the R-22 and water experiments significant; that is, can the low-temperature explosions be attributed to a low spontaneous nucleation temperature due to interfacial surface tension effects?

3. Is the interpretation of an interaction temperature valid; that is, at the end of a delay time, is the interaction initiated at the spontaneous nucleation temperature or at some threshold temperature at which film boiling becomes unstable?

4. Does a one-step, bulk-type interaction occur, as opposed to the propagating concept?

5. Is a pre-existing vapor phase a necessary condition for efficient intermixing?

6. What effect does surface solidification of the hot phase, for reactor-type materials, have on the explosion process?

7. Is hydrodynamic fragmentation greater or less for liquid-liquid systems compared with gas-liquid systems?

8. Which factors primarily influence fragmentation and intermixing?

Although an answer to each of the above questions is important to a basic understanding of the factors influencing vapor explosions, the question concerning the factors primarily governing fine-scale fuel fragmentation and intermixing with coolant is considered to be of primary importance,

since such phenomena form an integral part of the vapor explosion process. Indeed, it is generally accepted that fragmentation and intermixing are necessary conditions for explosive vaporization. As a result, the fragmentation process,
and more recently, intermixing considerations, have received
considerable attention in vapor explosion research. The
following section discusses these two interrelated phenomena.

V. FRAGMENTATION AND INTERMIXING CONSIDERATIONS

Although fragmentation and intermixing may not be sufficient in themselves to cause explosive vaporization, it is
clear that a large effective heat transfer area between fuel
and coolant must be generated in a rapid coherent manner if
explosive vaporization is to occur. A discussion of these
two important considerations is presented.

A.   Fragmentation

A rather extensive theoretical and experimental effort
has been devoted to understanding the fragmentation process.
Some of the more important modeling and experimental developments are reviewed, considering four general categories,
namely, those categories that consider either hydrodynamic,
boiling, internal pressurization, or solidification as governing effects. In addition, vapor blanketing effects also
are discussed in this section in the context of both fragmentation and intermixing. Models that have been discussed
in a previous paper(4) are covered only briefly here, whereas
the more recent developments and experiments are discussed
in greater detail.

1.   Hydrodynamic effects. If a molten droplet is subjected to velocity-induced surface forces sufficient to overcome the cohesive effects of surface tension, breakup of the
droplet may occur. The potential to cause hydrodynamic breakup
can be expressed in terms of the ratio of inertial-to-surface
tension forces, commonly called the Weber number.*

---

*   The Bond number, Bo (that is, the ratio of acceleration-
     to-surface tension forces) can be used also to assess
     hydrodynamic breakup, and is related to the Weber number
     as $3/4 \ C_D \ We = Bo$; where C_C is a drag coefficient.

$$We = \frac{\rho_c \, DV^2}{\sigma}$$

where

ρ_c = coolant density

D = diameter of droplet

V = velocity of droplet

σ = surface tension of droplet

In some of the earliest experiments on impact fragmentation performed by Ivins[60], low-melting-point metals (tin, lead, bismuth and mercury) were dropped from different heights into water at room temperature. Some of the results, shown in Figure 8, indicate that a fragmentation threshold occurs at a critical value between 10 and 20, which corresponds to that suggested by Hinze[61]. Cho[62] conducted similar experiments for the same materials while varying the quenching conditions and minimizing impact effects (that is, the Weber numbers were below the critical range). His work is summarized in Figure 9 which shows that thermal effects also influence the extent of fragmentation and may override those due to dynamic impact. This conclusion also is consistent with the recent experimental evidence of Lazarrus et al[63] which showed that extensive fine fragmentation of molten Al_2O_3 occurred when in an inert-gas atmosphere of argon at 2.5 torr. Such fragmentation was attributed to thermal stresses and will be discussed later.

Helmholtz (parallel to the contact plane) and Taylor (normal to the plane) instabilities also have been suggested as breakup mechanisms[64-67]. However, since the breakup of hot molten material often occurs in the absence of a significant velocity differential for dropping experiments[2,61, 62,68], in-pile fuel rod failure tests[69-71], and for true vapor explosions[1,2,7,16-19], this indicates that the fragmentation process is not likely to be controlled by hydrodynamic considerations alone,* although such effects may enhance breakup. Fine-scale fragmentation by thermal means appears likely[1,2,4,7,60,62,64,67,68,69].

---

* Hydrodynamic fragmentation may, however, be of primary importance to some situations; but foundary accidents and the S1-1, SPERT-ID, and BORAX-I explosions do not indicate this conclusively.

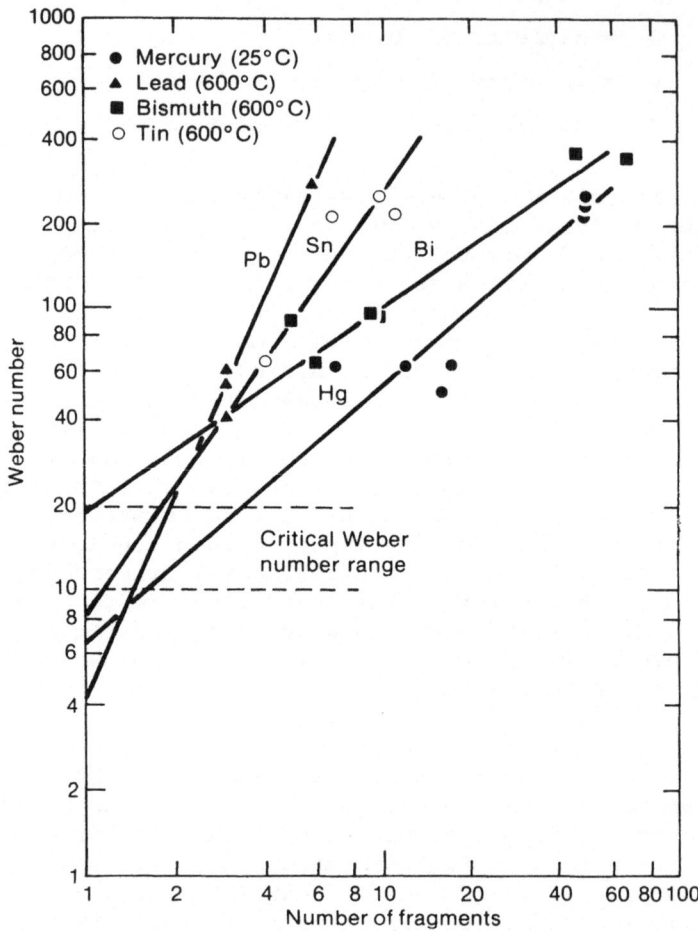

Figure 8. Plot of Weber Number Versus Number of Fragments
            for Molten Metals Dropped into Room-temperature
            Water (*60*)

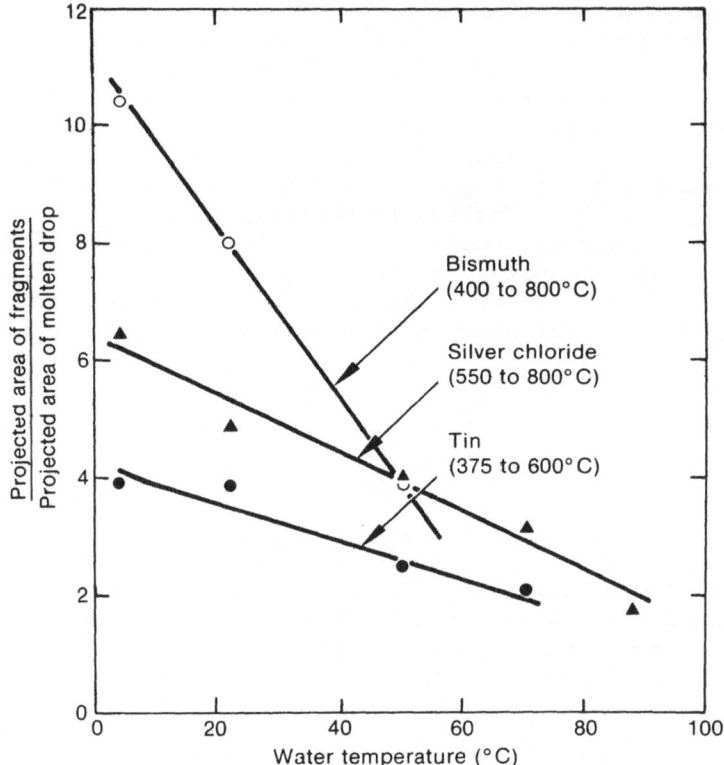

Figure 9   Fragmentation of Molten Metals Dropped into Water,
            Illustrating Effect of Coolant Subcooling(62)

        For jet-type fuel release from a failed rod, Weber
numbers on the order of 10^3 are expected, based on ejection
studies given in Reference 72 for a breach diameter of 0.25cm
and a velocity of 600cm/s. However, Bradley and Witte(73),
in experiments with mercury jets into water, have shown that
although some breakup occurs, it is much less than that for
heated jets; therefore, although the disruptive forces of
impact and viscous drag may contribute to breakup, it appears
that thermal effects play a more important role for high-
temperature materials.

        Several other authors(74-76) have considered var-
ious hydrodynamic-type fragmentation-intermixing models;

the most detailed with respect to MFCI phenomena is presented by Roberts(75). He considers that the interface between two liquids will increase exponentially with time
when one fluid is entrained in the other, due to the presence of spiral vortices (Figure 10) that result from an
'assumed' jetting process. Considering that surface and
friction forces can be neglected and that the relative
velocity spatial distribution is the same at all times, the
turbulent velocity is considered to be proportional to the
square root of the energy content of the vortex. Based on
this model concept, Roberts developed a set of equations for
the theoretical description of the increase in surface area
(a) and found that

$$A = \frac{A_o}{(1 - Bt \sqrt{A_o})^2} \tag{14}$$

where A is the initial interface area, and B is a constant
(s/cm), dependent on a number of scaling factors concerning
the kinematics of turbulence and associated energy dissipation processes. Equation (15) predicts an "explosion" of
the surface area and completion of energy release in a finite
time given as

$$t = \frac{1}{B\sqrt{A_o}} \tag{15}$$

Two obvious questions with respect to this model
are: (1), how the arbitrary constant, B, is to be assessed;
and (2), how the energy needed to initiate the vortices is
generated. In addition, the assumption of a vortex-type
geometry for the intermixing of two dissimilar liquids may
not be valid for the case where rapid heat transfer and phase
change occur, with vaporization of the cold fluid and solidification of the hot material. It is therefore felt that
the fragmentation-intermixing process of two liquids with
highly dissimilar thermal conditions is not primarily governed by the formation of such vortices, although such a
process may apply to the case of mixing two liquids at or
near thermal equilibrium. [*]

---

[*]   It is noted that if bubble growth and collapse are
      assumed, such collapse might initiate the formation
      of such a vortex(77).

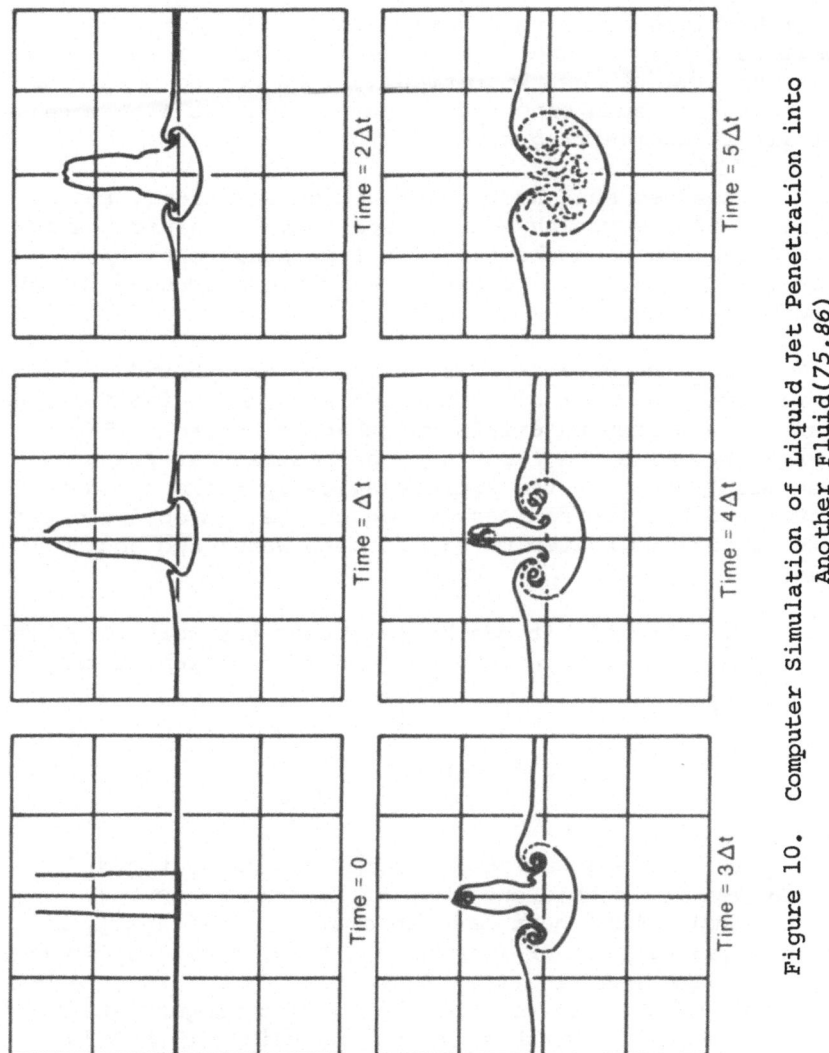

Figure 10. Computer Simulation of Liquid Jet Penetration into Another Fluid(75,86)

A somewhat similar turbulent mixing model has been developed by Bruckner and Unger(76). They coupled the kinetic energy of the turbulent field with the heat flux between melt and coolant, assuming a certain degree of efficiency for the conversion of heat into mechanical energy. However, similar problems, as with Roberts' model(75), exist with respect to assessing undetermined scaling constants. Until either experimental or theoretical values can be determined with respect to such constants, such a fragmentation model is of little engineering utility.

Besides such hydrodynamic considerations, the effects of bubble growth and collapse have received considerable attention as mechanisms for fine-scale breakup of molten fuel in coolant. These effects are discussed in the following section.

2.   Boiling effects. Vapor bubble growth and collapse have received considerable attention as a mechanism for fragmentation of molten materials quenched in coolant. Swift and Baker(68) were the first to hypothesize that fragmentation might occur in the hydrodynamically violent transition and nucleate boiling regimes, based on their dropping experiments of various molten materials into water and sodium coolant.

In general, it can be said that the work potential for fragmentation can be related to some fraction of the bubble energy

$$E_b = \frac{4}{3} \pi R_b^3 \, |\Delta P| \tag{16}$$

where R_b is the maximum bubble radius and ΔP is the difference between the bubble and ambient coolant pressures. Various studies have been made to evaluate this energy by assessing both R_b and ΔP from either theoretical considerations or experimental data. As mentioned, Reference 4 contains a description of some of the boiling-fragmentation models in connection with MFCI and analysis for LMFBR's. In the present discussion, primary emphasis is placed mainly on modeling and experimental work associated with water coolant.

As discussed in References 77-84, several variations of the bubble growth and collapse mechanism have been developed to describe fuel fragmentation. Usually, such modeling work has been performed in conjunction with small-scale, out-of-pile experiments with molten metal samples being dropped into water. Basically, such models consider that quiescent film boiling, if it occurs initially, is destabilized (owing to various hypothesized 'triggers'), resulting in bubble formation at the hot surface. During expansion, spherical geometry is assumed, such that bubble growth is described by the well-known Rayleigh equation (usually neglecting viscous and surface tension effects)

$$R \, \frac{d^2R}{dt^2} + \frac{3}{2} \left(\frac{dR}{dt}\right)^2 = (P_b - P_\infty)/\rho_c \qquad (17)$$

where

R = bubble radius

P_b = bubble pressure

P_∞ = ambient pressure

ρ_c = coolant density

Buchanan(77,78) assumes adiabatic expansion so that the mass of vapor in the bubble is constant, while Caldarola and Kastenberg(79) consider vapor addition. Following the growth stage, initiation of collapse is assumed when the bubble has penetrated into the cold surrounding coolant. This general description of the initial sequence of events is common to the models described in References 77-81; however, the assumed collapse and fragmentation mechanisms are somewhat different. Buchanan(77) considers that because of the presence of a molten heating surface, asymmetry of the collapsing bubble occurs, forming a high-velocity coolant jet(85) that penetrates the hot surface. Such coolant penetration into molten fuel results in turbulent mixing (similar to that described by Roberts(75) and Christiansen(86)), increased heat transfer area and the formation of a new bubble leading to eventual fragmentation by a cyclic process of bubble growth, collapse and jet penetration. Parametric calculations were carried out for the case of a fuel type considered to be seven times more dense than water coolant. Results

indicate that the ratio of pressures in successive cycles
depends on how the coolant is vaporized; that is, whether
by heterogeneous or homogeneous nucleation. For a system
external pressure of one bar, the cyclic pressure ratio is
calculated to be

$$\frac{P_i - P_o}{P_{i-1} - P_o} = \begin{cases} 6.67 \\ 2.90 \end{cases} \tag{18}$$

and the kinetic energy of the jet produced upon collapse is
approximately 0.44 times the energy of the ultimate bubble.
As illustrated in Table IV, the model predicts a buildup of
energy from a small perturbation to a rather energetic process, indicating that the model might be applicable to MFCI
analysis. As discussed in Reference 80 film studies(87,88)
with simulant materials show an interaction process that is
increasing with the time. Likewise, asymmetric vapor collapse in the form of jets has been noted experimentally(89).
Results of the model indicate that the jet energy, and thus
the fragmentation process, decreases with increasing system
pressure. Evidence of such an effect has been observed for
volcanic hydro-explosions where violent eruptions occur at
small sea depths(90). The consequence of this result is
that if the pressure of the initial cycle is not relieved,
a 'self-limited' MFCI will occur.* However, it is noted
here that since the entire vapor produced during each cycle
is assumed to collect in a single bubble, the model should
apply only to small masses. It is also felt that the rather
turbulent jet collapse and mixing process may lead to mul-
tiple-bubble formation with resultant energy density dissipation rather than an increasing energy density process,
which is a direct consequence of the single bubble concept.
It is suggested that visual experiments be conducted with
this critique in mind.

---

*   This would imply that increasing reactor system press-
     ure diminishes the probability of severe MFCI, which
     is a similar consequence of the spontaneous nucleation
     model, although based on different reasoning.

TABLE IV

ILLUSTRATIVE CALCULATIONS FOR BUCHANAN'S MODEL

ASSUMING A SYSTEM PRESSURE OF ONE BAR AND A HEAT TRANSFER COEFFICIENT OF 10^6 J/m² s K

Cycle (i)	Energy of Bubble E (i) (joules)	Final Radius of Bubble R_m (i) (cm)	Elapsed Time t (i) (s)	Mass of Coolant (kg)	Peak Pressure Difference at R_m, $P_i - P_o$ (bars)	Impulse at R_m (N/m²s)
1	1.261×10^{-4}	6.673×10^{-2}	2.687×10^{-3}	2.176×10^{-11}	3.339	5.354×10^{-2}
2	3.747×10^{-2}	4.453×10^{-1}	5.813×10^{-2}	6.487×10^{-9}	22.28	2.382×10^{-1}
3	$1.113 \times 10^{+1}$	2.971	1.197×10^{-2}	1.927×10^{-6}	148.7	1.060×10^{1}

Caldarola and associates(79,91,92) also have considered the bubble growth and collapse mechanism of fine fragmentation. The simplified Rayleigh equation again is solved; however, the process is considered to be nonadiabatic such that an assessment of heating effects are determined from Schlectendahl's analysis(93). The extent of bubble growth is calculated until inertia forces are overcome, at which point the bubble collapse begins. Knowing $R_{b,max}$ and P_b at this time, the maximum bubble energy is determined. Results for the case of UO_2 and sodium at various temperatures are presented in Table V. Although the bubble pressure is maximum near the initial stage, the radius is minimum; thus, during some expanded stage, bubble energy is maximized. For sodium, the maximum bubble energy is calculated to correspond to 52 atm pressure and a bubble radius of 3.14 cm. Since the temperature at this condition coincides with the minimum film boiling temperatures measured by Farahat(94,95) in his tantalum-sphere, sodium-coolant experiments, it was felt that the model might well explain fragmentation of UO_2 in Na (a similar analysis could be carried out for metal-water situations). However, it is noted that Farahat's experiments were carried out for solid tantalum spheres (1- to 1/2 inch diameter) with surface temperature conditions much lower than temperatures corresponding to molten UO_2; thus, his measurements of the minimum film boiling temperature may not accurately describe the actual situation for molten or solidified UO_2 at the small size of interest in MFCI analysis. It also should be noted that a rather large bubble size (3.14 cm) was predicted, which is greater than most experimental droplet sizes(31,60,62,64,69,96-99); thus, the vapor dome geometry assumed may not be realistic.

Caldarola and Kastenberg(79) also investigated the dynamics of bubble collapse in the form of microjets, illustrated in Figure 11. As a result of local impingement, an elastic wave is assumed to be generated in the molten fuel, with an associated acoustic energy (E_{ac}) per unit mass of

$$E_{ac} = \frac{1}{2}\left(\frac{\Delta P}{\rho_f C_o}\right)^2\left(\frac{R_j}{R_{ac}}\right)^2 \qquad (19)$$

where

R_{ac} = the distance of wave travel in the molten fuel

TABLE V

CALCULATION OF MAXIMUM BUBBLE RADIUS, $R_{b,max}$, PRESSURE DIFFERENCE BETWEEN BUBBLE AND COOLANT, ΔP, MAXIMUM BUBBLE WORK POTENTIAL, $W_{max} = 1\frac{1}{3}\pi R_{b,max}^3|\Delta P|$, AND THE APPROXIMATE ENERGY TRANSMITTED UPON BUBBLE COLLAPSE, E_{tr}, VERSUS SODIUM TEMPERATURE (79). IN ALL CASES, $E_{tr} = 0.15\ W_{max}$.

Sodium Temperature (K)	$R_{b,max}$ (cm)	$|\Delta P|$ (mPa)	W_{max} (J)
550	1.73	92.0	1.995
650	1.85	92.4	2.43
750	2.025	93.4	3.25
850	2.25	92.0	4.45
950	2.58	85.0	5.12
1050	3.24	61.8	8.02
1150	4.29	2.6	0.858

R_j = the jet radius(100)

C_o = the velocity of sound in the molten material

As opposed to the maximum potential energy of the vapor bubble, the acoustic energy associated with jet impingement is essentially independent of the bubble size. As shown in the last two columns of Table V, the energy deposition in the fuel due to collapse impingement is only a small fraction of the bubble work and much less than that required to account for fine-scale fragmentation. However, it is felt here that this estimate of the energy transmitted to the fuel is more realistic than that assuming 100% conversion of the maximum thermodynamic bubble work potential at the calculated departure radii. Thus, because of particle size to bubble radius considerations, it is felt here that the problem has yet to be treated in a realistic manner, and as such, the problem remains to define how much of the boiling energy is imparted to the fuel.*

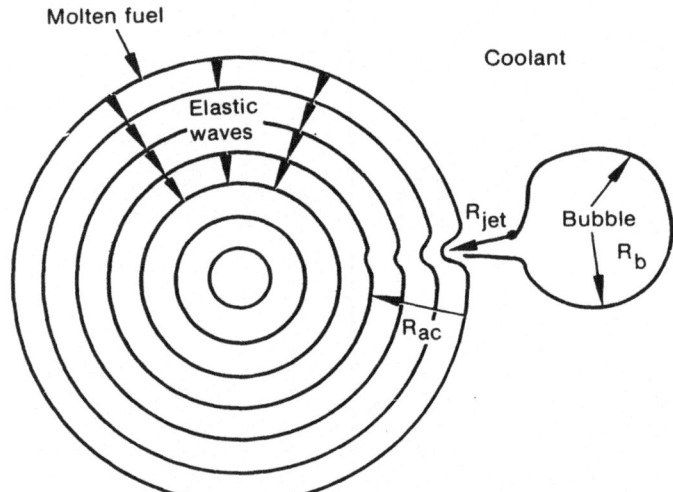

Figure 11.   Illustration of Jet Collapse Mechanism

---

*   It should be noted that bubble growth rate is maximum during the initial stages of inertia controlled growth (E); thus, the boiling forces imparted to the molten sample may be greater during the time of initial bubble growth rather than during collapse.

A somewhat different approach to the bubble growth and collapse concept, which considers the additional effect of surface solidification during quenching, has been formulated by Benz, Frohlich and Unger(*82-84*). Basically, the approach taken is to calculate the heat removal rate from the molten surface, assuming a nucleate boiling heat-transfer mode, where the bubble growth rate is calculated as

$$R_b(t) = 0.5 \left[0.234 \ \frac{k}{\rho L} \ \left(\frac{\rho \sigma}{\mu^2 \psi}\right)^{0.55} Pr^{0.33} \right]^{0.69}$$

$$(T_w - T_\infty)^{0.69} \ t^{0.69} \tag{20}$$

where

σ = surface energy

μ = viscosity

ψ = bubble detachment angle

T_w = surface temperature

Pr = Prandtl number

The other symbols are as previously defined. Equation (20) is based on Beer's(*101*) curve fit of experimental bubble growth rate data from a hot plate. Bubble collapse is considered to occur when the inertial forces are overcome. Assuming that some fraction (η) of the maximum bubble potential energy is used in the creation of new surface area upon bubble collapse, the increase in surface area (F_s) can be expressed as

$$F_s = \frac{4 \pi \eta (\Delta P) R_b^3}{3 \sigma} \tag{21}$$

An iterative calculation is made for each bubble growth-collapse period, during which time an assessment is made of the heat transfer process associated with solidification of the molten material. As illustrated in Figure 12, such a process is exponentially increasing with respect to surface area generation (F_s/F_o) and exponentially decreasing with

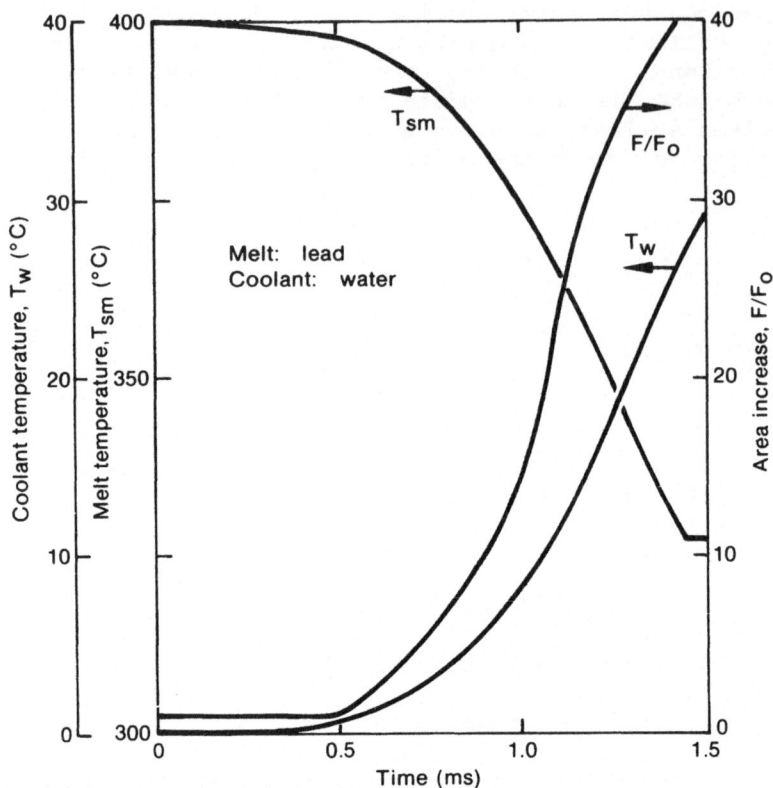

Figure 12.   Illustration of Surface Area Generation (F/F_o),
Course of Melt Temperature (T_{sm}), and Water
Coolant Temperature (T_w) as a Function of
Time(82)

respect to the solidification process (illustrated as T_{sm}).
Such a model, however, is somewhat parametric in nature,
due to the necessity of assessing such parameters as η and
fraction of heat flux for surface solidification. Comparison of the initial and final surface areas in controlled
dropping experiments may, however, lead to an evaluation of
such unknown parameters. An attempt at such an evaluation
presently is being pursued, using Frohlich's experimental
results(102). The basic assumptions of the model, however,
are supported by several investigations(41,88,103) where

the extent of fragmentation correlates with the onset temperature for transition boiling and the fragmentation is
accompanied by solidification.

         In addition to both hydrodynamic and boiling
effects on the fragmentation process, internal pressurization within molten fuel, due to various sources, also has
been considered as a potential initiator of fragmentation.
A description of the various internal pressurization concepts is discussed in the following section.

   3.   Internal pressurization effects.

         Several mechanisms have been proposed that con-
sider coolant encapsulation by the hot phase as the principal cause of fragmentation. Long's(2) early experiments
and those of Hess and Brondyke(104) indicated that small
amounts of cold liquid were confined by the hot phase, resulting in an explosive molten aluminum-water interaction.
Similar results were reported by Sallack(105) with molten
smelt and water and by Brauer et al(106) with various molten metal-water systems. Sallack(105), however, considers
that a solid shell may form during encapsulation, with
shrinkage having an additional stress effect.

         Based on such experimental observations, Schins(107)
proposed a sequence of events that might lead to such encapsulation and fragmentation. Although a mechanistic description of events is proposed, no quantitative analysis of
the fragmentation process is given. The hypothesized sequence of events is depicted in Figure 13, and can be described sequentially as:

         a.   direct liquid-liquid contact,
               resulting in a rapid temperature
               increase of the adhering coolant
               layer

         b.   incipience of transition boiling,
               which imparts a shock to the
               molten fuel surface

         c.   collapse of the vapor film, which
               initiates cavitation of bubbles
               within the molten droplet

  d.   entrainment of coolant in such
       cavitated fuel

  e.   fragmentation caused by explosive
       vaporization of entrained droplets

       Although such a sequence of events may occur, the
question remains whether the process is sufficiently energetic to cause fine, coherent fragmentation, and whether
the kinetics of events simulate experimental results.   Until
a quantitative model development has been made, such questions cannot be answered.

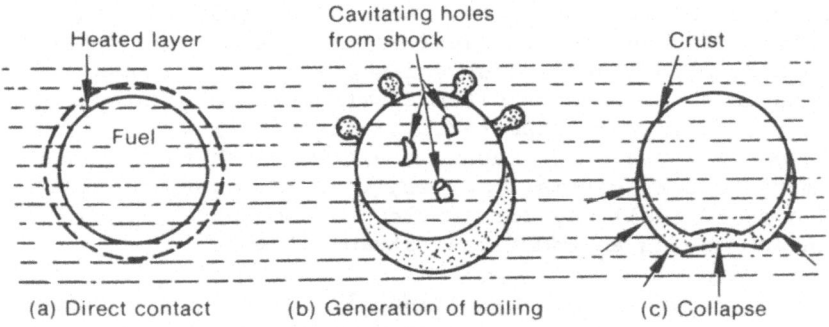

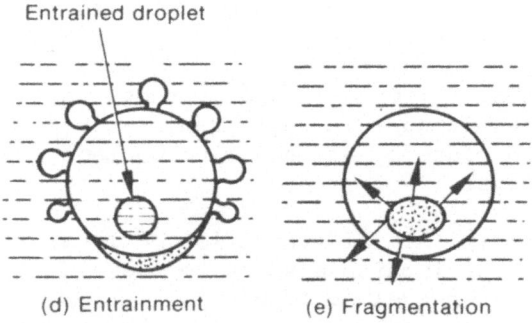

Figure 13.   Descriptive Illustration of the Schins
             Boiling Model for Fragmentation(107)

Besides the model concept of Schins, Kazimit(108) proposed a cavitation-induced fragmentation process as depicted in Figure 14. As illustrated, internal acoustic cavitation within the molten material is assumed to be induced by fluctuating pressure waves generated in the melt as a result of surface boiling and attendant film collapse. Two conditions necessary for fracture by cavitation are bubble inception and continued growth. In Reference 109, the minimum negative threshold pressure (P_{th}) for homogeneous cavitation is given as

$$P_{th} = \left[\frac{9.06 \ \sigma^3/kT}{\ln \dfrac{1.45 \ \rho N^2 \sigma^2}{P_{th} \ M^{3/2}RT} - \dfrac{L}{kT}} \right]^{1/2} \tag{22}$$

where

k = Boltzmann's constant

N = Advogadro's number

M = molecular weight

L = latent heat of vaporization

R = gas constant

The other symbols are as previously defined. This condition was not satisfied in parametric calculations performed by Kazimi(108). However, it was postulated(108) that, due to the presence of impurities, cavitation may occur in laboratory experiments or under reactor conditions. An alternate criterion, suggested in Reference 110, is to consider the presence of small amounts of gas in the molten droplet. For this condition, the threshold pressure can be calculated, based on the assumption of equilibrium between pressure and surface tension forces, and is given by(111)

$$P_{th} = \Delta P + \frac{4\sigma}{3 \ \sqrt{3} \ R_{gn}} \left(1 + \Delta P \ \frac{R_{gn}}{2\sigma} \right)^{-1/2} \tag{23}$$

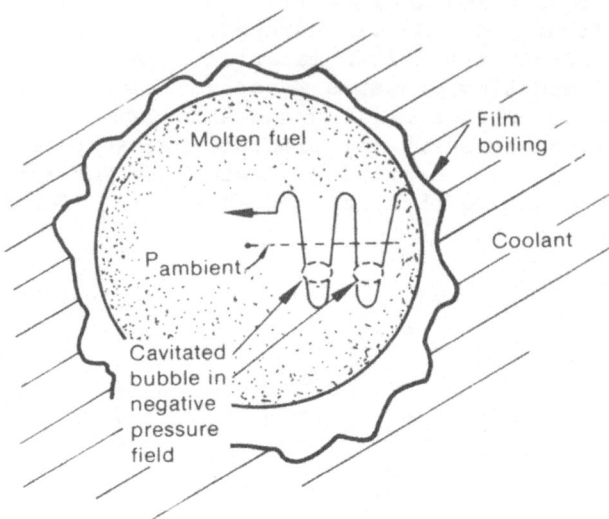

Figure 14.   Descriptive Illustration of the Acoustic
             Cavitation Model for Fragmentation

where

ΔP = difference between system and vapor pressure

R_{gn} = radius of the gas nuclei $(2.0 \times 10^{-5}\text{cm})$ (111)

        For molten UO_2 at a system pressure of 1 atm, P_{th}
is approximately 20 atm, which is about the order of magnitude of the internal pressure amplitudes calculated by
Kazimi(108). Thus, such cavitation may occur in molten UO_2
if extraneous gas is present. However, even if such a
cavitated bubble could nucleate, it must survive and grow
in a fluctuating pressure field such that the rate of growth
must be greater than the rate of collapse to achieve a sufficient buildup of bubble energy to cause fragmentation.
Since no assessment of either the work potential to cause
fragmentation or the growth kinetics was made, the validity
of the model is difficult to assess, other than to say it
does not satisfy homogeneous cavitation requirements.

Other general variations of the internal pressurization concept are the violent gas release model proposed by Epstein(*112,113*) and the impulse-initiated gas release mechanism of Buxton and Nelson(*114,115*). As illustrated in Figure 15, Epstein's model assumes that dissolved gases are present in the molten sample. During rapid quenching, it is assumed that the liquid becomes supersaturated, such that violent gas release causes fragmentation. To account for breakup by this mechanism requires that the melt be capable of dissolving gas without forming a stable phase and that the solubility decreases substantially at quenching temperatures. Although some low melting point materials may exhibit favorable solubility characteristics and disruption of molten steel (due to deoxygenation) has been observed experimentally(*116*), most fragmentation metal-water experiments have been conducted in an inert atmosphere (He, Ar) where relatively low solubilities can be expected(*117*); yet extensive fragmentation still occurred. With respect to UO_2, it also has been demonstrated by Gunnerson and Cronenberg(*118*) that the gases present in a reactor environment (for example, He-bond; Ar-cover gas; Xe and Kr fission products), solubility characteristics would not favor such a fragmentation mechanism.

Buxton and Nelson(*114*) also proposed a variation of such a gas release mechanism by considering the internal bubble nucleation process to be impulse-initiated. The principal characteristics of the model are:

    a.   achievement of a large quantity of dissolved gas in the molten phase

    b.   supersaturation of such dissolved gas as the melt is quenched by surface coolant boiling

    c.   bubble nucleation by an applied impulse trigger

    d.   rapid bubble growth resulting in fragmentation

The concept that bubble nucleation can be impulse-initiated certainly helps trigger such a gas-release phenomenon. Impurities in the melt also will help initiate a

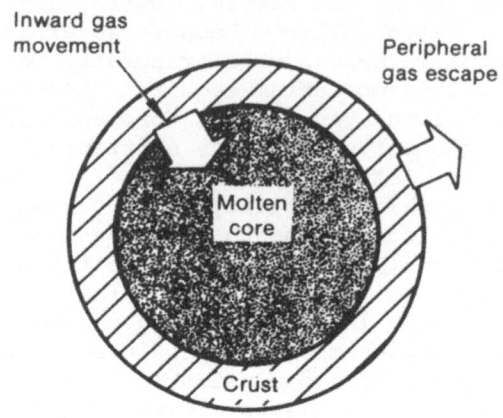

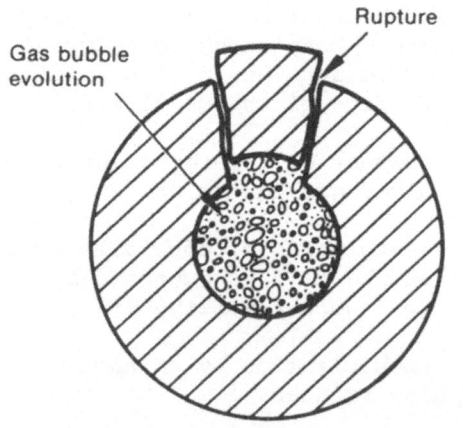

Figure 15.   Illustration of Gas Evolution Model(*113*)

bubble nucleation process. However, many experiments have
been conducted in inert or evacuated environments with short
heating times, in the absence of known impulse triggers, yet
extensive fragmentation occurred. Likewise, since solubility characteristics are primarily controlled by the
chemical nature of the gas-melt system and vary accordingly,
a marked difference in the extent of fragmentation should
be expected in accordance with solubility trends; however,
as listed in Reference 7, the large body of fragmentation
experiments does not lead to such a correlation, but rather,
leads to the conclusion that thermal versus chemical effects
appear to govern fragmentation. Thus, although solubility
and surface oxidation(116,119) effects may contribute to
breakup,* they are not considered here to be controlling
factors that govern a wide range of metal-water or oxide,
fuel-sodium systems.

4.   Solidification effects. In the models described
previously, it is generally assumed that prior to and during
fragmentation, the quenched materials remain in the molten
state. However, several experimenters(63,67,69,106,120)
have noted that rapid quenching leads to surface solidification. Such solidification may result in a thermal, stress-
initiated fragmentation process. Variations of thermal
stress models are given in References 121-123.

To assess in a quantitative manner the potential
for fragmentation by shell solidification, Hsiao et al(121)
analyzed the case of initially molten aluminum quenched in
water. In Reference 122 such analysis was extended, considering the effects of temperature-dependent properties,
compressibility of the inner molten core, various modes of
surface heat transfer, and comparison with a UO_2-Na system.
In Reference 122, it was found that the assumed heat transfer condition is of primary influence. For the Al-H_2O case,
assuming perfect contact, the generated surface tensile
stress was estimated to exceed the yield strength, demon-
strating that surface rupture can be expected. However, for

---

* It should be noted that surface oxidation also may
   influence the heat transfer process for particular
   systems (for example, copper quenched in water(120)),
   which, in turn, may influence the extent of fragmentation.

a film boiling heat transfer mode, surface fracture is not predicted. Qualitatively, this compares favorably with experiments in which molten aluminum was forcibly injected into water, with the probability of metal-coolant interaction and consequent fragmentation(124), whereas dropping experiments with initial film boiling produced little breakup(68). For UO2 fuel (with a much lower conductivity, and thus a larger surface temperature gradient), the thermal stresses are considerably higher(122) for equivalent quenching conditions. The effects of thermal, shock-induced crack generation have been studied by Knapp and Todreas(123). It was again demonstrated that rapidly quenched ceramic fuels, such as UO_2, would lead to crack formation; thus, the possibility for coolant encapsulation and further fragmentation.

Zyszkowski(120,125) also considers that thermal interaction for copper-water systems is influenced by the solidification process of the molten copper. According to his hypothesis, an interaction will occur when molten metal solidifies and some undefined internal mechanism in the metal causes fragmentation. Such a concept is based on experimental results where a thermocouple that penetrated the copper drop at the time it reached the base of the container vessel, indicated a temperature of about 1300°C prior to explosive-type vaporization. However, violent interactions occurred only when the molten copper was heated in air, forming an oxide layer. Such an oxide layer may have a destabilizing effect on film boiling, thus influencing the interaction process. To test Zyszkowski's hypothesis further, a wide range of materials with very different melting temperatures would be needed. However, since the fuel surface temperature at the time of interaction is an important parameter in any discussion of MFCI events, his experimental technique is most pertinent.

The fact that the thermal energy is directly deposited in the form of stress within the hot phase, versus the boiling models where a significant fraction of the energy is imparted to the coolant upon bubble collapse, lends support to the stress-induced fragmentation concept. Likewise, some experimental evidence(63,69) exists for such a mechanism for ceramics; however, most metals undergo plastic rather than brittle deformation so that many of the metal-water fragmentation experiments cannot be accounted for by such thermal stress models. In addition, an order of magnitude comparison(51) of one-step mixing energy requirements ($E_{mix} \simeq 3\rho V^2 / 8t^2$ R; where V = initial volume, t = mixing time, and R = fragmented radius) with the elastic energy stored in a thermally-stressed shell ($E_e \simeq \sigma_f^2$ V/2Y; where σ_f = fracture stress and Y = Young's modulus) indicates that intermixing cannot be accounted for by the stress mechanism alone.* However, nucleate boiling with attendant rapid heat transfer may lead to both sufficient thermal stress and vapor production, such that the latter effect can sufficiently reduce the mixing energy requirements; thus, thermal stress may initiate breakup, whereas vapor production leads to intermixing.

One of the principal concerns of the shell solidification concept is whether or not crystallization occurs at the quenching surface for the times of interest in MFCI analysis (that is, on the order of several ms). In References 49 and 50 various aspects of crystallization kinetics (that is, the rate of solid crystal nucleation and the rate of growth of solid in its melt) were investigated and compared with the maximum rate of solidification heat transfer for good fuel-coolant contact. Results indicate that for good quenching conditions, surface crystallization of UO_2 commences almost immediately (within 1 or 2 ms). This is of importance not only to fragmentation analysis, but also to overall modeling of phase transformation kinetics, particularly in the context of the spontaneous nucleation model where Volmer's theory is used to describe phase change kinetics in the coolant but has not, to date, been considered with respect to solidification of the fuel in the overall model.

From the discussion of fragmentation models, it can be generally summarized that if either coolant vapor or inert gases blanket the fuel surface, the fragmentation process will be diminished. This has been demonstrated in many experiments where fragmentation is quite limited when various molten materials are quenched in the film boiling

---

*   The concern of energy requirements for intermixing is not limited to the solidification concepts alone, but also applies to the fragmentation models previously discussed.

regime. From a fragmentation standpoint alone, it therefore would appear that the presence of noncondensable gases
or vapor decreases the probability of an energetic interaction. However, Cho(45,46), in his work on intermixing
considerations for vapor explosions, indicated that the
presence of vapor or noncondensable gases may actually increase the chances for such violent MFCI-induced vapor explosions. A brief discussion of intermixing considerations
is presented in the following section.

B.   Intermixing

     With respect to breakup and intermixing, Cho et al(45,
46) consider that such processes are governed by frictional
energy dissipation, with a geometric progression of breakup
and a constant mixing velocity during each stage. Based on
these assumptions, the minimum mixing energy is found to be

$$E_{mix,min} = 1.81 \, \rho \, V_f \left(\frac{V_f^{2/3}}{t_m^2} \right) \left(1 - \frac{R^2}{V_f^{2/3}} \right) \ln \left(\frac{V_f^{1/3}}{R} \right) \quad (24)$$

where

ρ = coolant density

V_f = initial volume

t_m = mixing time

R = fragmented particle size

     If this mixing energy is less than some fraction of
excess thermal energy of the fuel ($E_f = \rho \, C_p \, V_f \, \Delta T$; where
ΔT is the temperature of the fuel above that of the coolant),
it can be said that the energy requirements for rapid, fine-
scale intermixing are satisfied.

     As illustrated in Figure 16, two different, one-dimen-
sional cases are considered by Cho et al(45). For a coarse
mixture of molten fuel and coolant in the absence of vapor
(case a), the two fluids cannot begin to accelerate effectively (or mix) until the disturbing force is relieved at
the end of the fuel-coolant column. However, if the fuel

and coolant layers are separated by a compressible vapor
(case b), then they can accelerate immediately after disturbance relief in the vapor region. Consideration of these
two situations indicates that the existence of a compressible
layer reduces the effective mass to be accelerated, thus resulting in a much finer "localized" intermixing. This suggests that film boiling prior to a pressurization would result in a more effective intermixing process which, in turn,
could result in violent explosive interaction. An examination
of the SPERT-ID test(19) supports such arguments. The results of such experiments indicate that a vapor explosion is
accompanied by a fragmentation process, resulting in particle
sizes of approximately 100 m in radius (R) and a mixing time
(t_m) of a few milliseconds or less. If a mixing time of
5 ms is assumed, and if it is assumed that 1% of the excess
fuel sensible heat is used in the intermixing process, the
energies can be compared. As illustrated in Table VI, the
volume of each of the fuel elements for the SPERT-ID(19),
BORAX-I(17), and SL-1(18,126) reactors are small enough so
that intermixing energy requirements necessary for a vapor
explosion are satisfied.

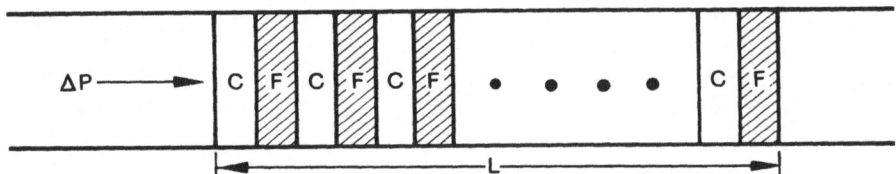

(Case a) No vapor layer between fuel (F) and coolant (C) layers.

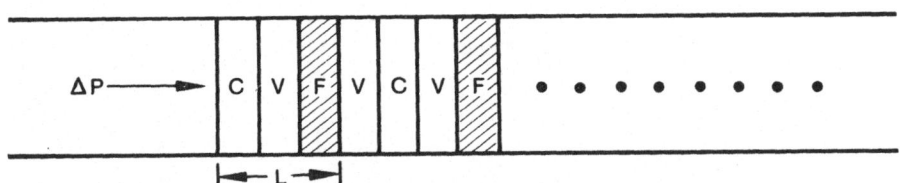

(Case b) Vapor (V) layer between fuel (F) and coolant (C) layers.

ΔP = pressure differential
L = reaction zone length

Figure 16.   Illustration of Fuel-Liquid and Fuel-Vapor-
              Liquid Mixing Geometries

TABLE VI

COMPARISON OF MIXING ENERGY REQUIREMENTS WITH AVAILABLE SENSIBLE HEAT OF FUEL
AT TIME OF VAPOR EXPLOSION IN THREE DIFFERENT WATER-COOLED REACTORS

	SPERT-ID	BORAX	SL-1
Considered cause of reactor damage	Vapor Explosion	Vapor Explosion	Vapor Explosion
Fuel element design	Plate	Plate	Plate
Fuel dimensions in active core	0.02 x 2.45 x 23.5 in.	0.021 x 2.5 x 2.36 in.	0.05 x 3-7/8 x 26 in.
Fuel volume	1.15 in. (18.25 cm)	1.24 in. (20.3 cm)	5.04 in. (82.55 cm)
Fuel type	U-Al alloy	U-Al alloy	U-Al alloy
$E_{min,max}$ (Equation 4) R = 100 μ (0.01 cm) t = 2 ms	8.05 cal	9.16 cal	102.9 cal

	SPERT-ID	BORAX	SL-1
ρ_f		~3.10 g/cm	
C_p, f		~0.225 cal/g °C	
T_{melt}, f		~940°C	
T		~900°C	
E_f (at 1%)	1.18 cal	131 cal	518 cal
Comparison		E_f (at 1%) $E_{mix,min}$	
Potential for vapor explosion	Yes	Yes	Yes

Although such calculations only can be considered first-order estimates, the fuel rod bundle concept appears to satisfy the initial dispersal requirements for vapor explosions. A more detailed calculation might include kinetic energy requirements for fuel particle movement through the coolant and fragmentation energy requirements considering such phenomena as bubble growth and collapse or thermal-stress-induced breakup.

VI. NUCLEAR REACTOR CONSIDERATIONS

As indicated from the previous discussion, most recent explosion research has been centered primarily on obtaining an understanding of the necessary conditions for explosive vaporization. Although, in a general sense, some consensus of opinion is emerging on defining such necessary conditions, there appears to be little agreement on the details of the underlying phenomena. Thus, any discussion of vapor explosion potential in nuclear reactor systems is subject to a somewhat limited knowledge of the problem. Nevertheless, some general remarks can be made concerning accident initiators and preventive measures. It should be noted, however, that present knowledge indicates that the necessary conditions appear to be difficult to meet in reactor systems, and because of redundant safety features, vapor explosions have never occurred in commercial power reactors; their probability of occurrence is low(8).

A.   Accident Initiators

Initially, severe transient overpower (TOP) accidents were of primary interest relating to the MFCI vapor-explosion problem as applied to LWR systems, where such accidents must have a sufficiently high reactivity rate (several $/s) to cause fuel melting and rod failure prior to significant voiding of liquid coolant from the core. Indeed, as discussed, the SPERT-ID and BORAX-I experiments and the SL-1 accident were of this type. However, with present-day power reactor designs calling for low reactivity worth of both individual fuel bundles and control rods, the probability of initiating such reactivity excursion accidents is insignificant. Various loss-of-flow (LOF) scenarios can be hypothesized, resulting in the potential for MFCI-induced ex-

Thus it would appear that the probability of such an event is somewhat less likely for irradiated fuel than for fresh elements, due to the presence of gaseous fission products that are released from the fuel matrix. Trends of the TREAT fuel experiments(128-133) indicate such an effect, although none of these experiments underwent a true vapor explosion, but rather milder MFCI-induced coolant voiding. Likewise, fuel element designs calling for a gaseous bond such as He would appear to be safer than concepts calling for metallic bonded fuel elements or sandwich-type designs used for low-power LWR's (for example, fuel clad directly to an outer metallic sheath, similar in design to the SPERT-ID, BORAX-I, or SL-1 fuel designs). For similar reasons, vented fuel elements would not appear to be safer than designs calling for a gaseous plenum region within the fuel element itself, although such vented elements might be more desirable when considering normal swelling behavior or less severe off-normal events such as localized overheating. In addition, single-phase liquid coolant reflooding, approaching a hammer-type situation, with a partially molten core, should be avoided.

With respect to the spontaneous nucleation concept, several factors might be considered that would diminish the probability of explosive vaporization, if such a concept actually applies to reactor systems. As pointed out by Fauske(29), the presence of radiation, gaseous fission products, solid particulate matter and other nucleation aids tends to cause heterogeneous nucleation prior to reaching the temperature limit at which rapid change-of-phase would occur, thus initiating mild MFCI rather than a true vapor explosion.

With respect to the pressure detonation model, elimination of an initial trigger pressure pulse essentially would eliminate the possibility of a vapor explosion based on this model concept; thus, the presence of noncondensables would again be advantageous. Also, designs that eliminate rigid, long-tube geometries would be desirable.

VII.   SUMMARY

As a result of knowledge gained from recent research efforts, it can be summarized that a vapor explosion is characterized by the following factors: *

1.   a period of stable film boiling

2.   destabilization of film boiling by either thermal or pressure-induced means, or both

3.   intimate contact between molten fuel and coolant

4.   extensive fuel fragmentation and inter-mixing with liquid coolant, resulting in a large, effective heat transfer area, causing rapid coherent coolant vaporization

5.   sufficient system constraint to cause severe shock pressurization

At the present time, however, there appears to be little consensus of opinion as to the details of the actual physical phenomena involved in such processes. For example, there appears to be little agreement on the actual mechanism of fragmentation, the rate at which it occurs, and the ultimate particle size that can be expected. Likewise, the question of the energy transfer mechanism between fuel and coolant is still unanswered. On one hand it has been proposed that rapid phase-transformation by spontaneous nucleation is a necessary condition for obtaining a vapor explosion, whereas others have postulated that a nucleate-type boiling process from finely-divided fuel intermixed with coolant can produce vapor equally at a sufficient rate to obtain shock pressurization of the system. In spite of such uncertainties, it appears that some of the basic phenomena associated with vapor explosions can be characterized at this time. As fundamental research continues on

---

*   The first two factors may be inappropriate for "water-hammer" impact-type conditions.

the subject, a reevaluation of the pertinent phenomena may
be necessary. Until such time, the following features are
suggested for incorporation into simplified parametric calculations.

A.   Fragmentation-Intermixing Process

As discussed previously, current research efforts have
been centered primarily on a determination of the principal
mechanisms involved in fragmentation, rather than on an
assessment of the kinetics of such breakup and the resultant
particle size distribution that can be expected for a defined set of initial conditions. However, both in-pile
fuel-failure tests and out-of-pile molten metal dropping
experiments indicate that the process is primarily thermal
in nature, with hydrodynamic influences having an added
effect. Nevertheless, because no one model yet has been
accepted that enables the user to estimate the rate of breakup and expected particle sizes, one must resort to empirical
evidence for parametric calculation.

Mizuta's(69) correlation for the particle size distribution appears appropriate, since it is based on an extensive compilation of data from simulant materials, as well
as UO_2 fuel quenched in both sodium and water. Such data
include in-reactor UO_2-Na and UO_2-H_2O results, laboratory
induction-heated and thermite-reaction UO_2 into Na, and data
from the SPERT-ID destructive test. The particle size distributions from such experiments show remarkable consistency
and normal Gaussian distribution when displayed on a logar-
ithmic scale. Mizuta's distribution function for the fines
is

$$f\ (\log D) = 58.1 \exp \frac{-(\log D - 2.27)^2}{0.944} \qquad (25)$$

where D is the particle diameter (μm), the most probable
mean-mass size being $\simeq$ 200 μm, with a maximum bound of
$\simeq$ 2500 m. It is cautioned that this correlation is based
primarily on fragmentation data obtained from controlled,
small-scale experiments (some of which cannot be considered
vapor explosion events), rather than the situation of a
large-scale vapor explosion for a nuclear reactor; thus,
such a correlation is subject to questions of prototypicality

The large-scale UO_2-Na French experiments(134,135) should help in clarifying such uncertainties.

As discussed by Cho(44,45), an appropriate breakup and intermixing time, based on the results of parametric calculations(46) and the results of the SPERT-ID(19) destructive experiments, is on the order of a few milliseconds. Further evidence that such mixing and fragmentation occurs on a millisecond time scale can be found in numerous out-of-pile dropping experiments and also in the analysis of the H-2 in-pile fuel failure experiment(132,133) where the total time for cladding expansion and rupture, fuel release from the interior of the rod, breakup, and intermixing, was found to be no greater than 40 ms. A time for breakup and intermixing alone of several milliseconds, therefore, is suggested.

B.   Heat Transfer-Pressurization Process

Besides the condition of a rapid fragmentation-intermixing process, it is generally agreed that intimate fuel-coolant contact occurs, resulting in rapid vaporization and shock pressurization. However, there is little agreement with respect to the mechanics involved in the heat transfer-vaporization process. On one hand, several authors(29,35) consider a heat transfer-vaporization process, asserting that if the instantaneous contact temperature between a large mass of fuel and coolant exceeds the spontaneous vapor nucleation temperature, a vapor explosion will occur. Others(33,38,52,53,75,77,83) consider a progressive process wherein continued fragmentation/heat transfer-pressurization can lead to explosive vaporization. With respect to the spontaneous nucleation model, it has been argued that a vapor explosion can occur for a fuel-water system, but not for the large-mass fuel-sodium case (because the instantaneous contact temperature is below that for homogeneous nucleation of sodium). The progressive models, however, predict that vapor explosions can occur for many fluid systems (including UO_2-Na) under favorable hydrodynamic conditions (which require a rather detailed analysis). For this reason, present research efforts have concentrated on identifying the basic criteria and mechanisms involved. Until such issues are settled, a unified heat transfer-vaporization-pressurization model, which is capable of predicting the important state variables necessary for estimating work

potential (that is, specific volume and pressure), cannot
be established with any degree of confidence at this time.

VIII.   CONCLUSIONS

At the time a previous review paper(1) was published
in 1973, very little in the way of overall modeling of the
vapor explosion problem existed, other than the simplified
equilibrium thermodynamic approach of Hicks and Menzies(27)
and the liquid thermal expansion concept of Cho and Wright
(136,139). Neither of these approaches, however, addressed
the problem of assessing the criteria for, and transient
phenomena involved in, rapid vaporization-induced shock
pressurization (that is, a true vapor explosion). Since
that time, several overall mechanistic type models have
been developed that attempt to describe the actual mechanisms involved and the kinetics of such processes, the most
significant concepts being the spontaneous nucleation approach of Fauske(29,30) and Henry(35-37), and the pressure
detonation model of Board and Hall(38,52). Although each
of these approaches has its supporters, neither has gained
wide acceptance in the engineering community as an acceptable description of the actual phenomena involved or as
encompassing all possible scenarios that could lead to explosive vaporization. It is the opinion here that the
spontaneous nucleation model is one mechanism by which
vapor explosions can be accounted for; however, it alone
may not be considered the basic criterion upon which vapor
explosion potential should be judged. An 'a priori' conclusion that a vapor explosion will not occur in a reactor
system if the instantaneous contact temperature between fuel
and coolant is below the spontaneous nucleation temperature
or if the system pressure is greater than the critical pressure of the working fluid, cannot suffice if it can be demonstrated that progressive fragmentation-heat transfer-pressurization models can lead to explosive vaporization. The
work of Board and Hall is an attempt at such an approach;
however, as discussed previously, serious questions arise
concerning the original hydrodynamic assumptions and associated fragmentation dynamics; thus, further investigation
of this concept appears warranted. The work of Anderson and
Armstrong(33), Sharon and Bankoff(59), Patel and Theofanous(5
and Williams(58) should help clarify some of these questions.

The fact that extensive investigations with simulant
fluid experiments appear to support the interface-spontaneous nucleation concept indicates a cautious validity of
the arguments presented by Fauske and Henry. However, it
is emphasized that most simulant fluid experiments have
been carried out to test particular model assumptions
rather than to assess all-encompassing conditions that
could lead to shock-type explosive vaporization. For example, the water-hammer experiments of Wright and Humber-
stone(71) produced vapor explosions with molten aluminum
impacted upon water (which satisfies the contact interface-
nucleation criterion); such experiments have not been conducted for molten UO_2 impacted upon sodium (which would not
satisfy the contact interface-nucleation criterion), yet
explosive vaporization under such conditions appears possible. Therefore, $T_I > T_{SN}$ is not considered to be a necessary condition for explosive vaporization, but rather, just
one mechanism by which rapid vaporization can be accounted
for. Further experimental verification of required conditions is thus deemed necessary before either the pressure
detonation concept or other plausible scenarios can be completely discounted in reactor safety assessment.

An overview of various detonation concepts and the
spontaneous nucleation model is illustrated in Table VII,
and indicates that some commonality of thought exists in
that both attempt to attribute some necessary trigger criterion to account for explosive vaporization, the former
requiring pressure-induced or thermal-induced film boiling
destabilization and fragmentation, the latter requiring a
quantum jump in vapor nucleation rate at a temperature condition considered to be the spontaneous nucleation temperature. Considering the ideas presented in these model
concepts and experimental results, it appears to the authors
that destabilization of quiescent film boiling, due to either
pressure or thermal conditions or both, resulting in rather
violent nucleate-type boiling and fragmentation into a large
surface heat transfer area, satisfies the conditions for explosive vaporization. The fact that the predicted homogeneous nucleation temperature is not all that different from
a thermodynamic prediction of the minimum film boiling tem-
perature(39,43,138) (Leidenfrost point) and that contact-
wetting conditions affect both, indicates that experiments
that have been interpreted in light of a spontaneous or
homogeneous nucleation temperature(35,37) may also be ex-

TABLE VII

COMPARISON OF VAPOR EXPLOSION CONDITIONS

Vapor Explosion Conditions	Fauske-Henry(1,2)	Board-Hall(3)	Anderson-Armstrong(4)	Cronenberg-Gunnerson(5,6)
1. Initially stable film boiling, so that vapor film separates the two liquids and permits coarse premixing without excessive energy transfer	Consistent	Consistent	Consistent	Consistent with all model concepts
2. Breakdown of film boiling	Due to thermal or pressure effects	Due to pressure effects	Due to pressure effects	Due to thermal effects
3. Fuel-coolant contact upon breakdown of film	Liquid-liquid contact	Liquid-liquid contact	Liquid-liquid,	Liquid-liquid or solid crust-liquid contact
4. Rapid vapor production, causing shock-pressurization	Due to spontaneous vapor bubble nucleation (assessed from kinetic theory) and fine-scale fragmentation-intermixing	Due to a large, effective heat transfer surface as a result of fine-scale fragmentation and intermixing	Due to a large, effective heat transfer surface as a result of fine-scale fragmentation and intermixing	Due to a large, effective heat transfer surface as a result of fine-scale fragmentation and intermixing
5. Adequate physical and inertial constraints to sustain a shock wave	Consistent	Consistent	Consistent	Consistent

plosive vaporization. Such LOF accidents could lead to
fuel overheating and failure, where subsequent reflooding
of the reactor core might result in coolant being forced
into contact with molten fuel. If such intimate contact
between molten fuel and coolant were to occur, the potential
for a vapor explosion could exist (as demonstrated by
Wright's water-hammer-type experiments with molten alumin-
um(71)).

     With respect to LMFBR's, both TOP (because of a poten-
tial positive void coefficient) and LOF accidents are of
interest. In addition, the core of a large, fast reactor
is not in its most critical configuration, such that there
is the added concern of a recurring criticality and coolant
reentry problem(127), which could lead to an explosive MFCI
situation. However, as discussed previously, the conditions
for the occurrence of a true vapor explosion in either reactor system appear difficult to meet, based on the present
understanding of phenomena associated with thermal vapor
explosions. Likewise, knowledge gained to date suggests
that certain reactor design steps might be given serious
consideration as preventive measures. Indeed, the present
practice of limiting single control rod worth below 1$, so
as to not cause a prompt critical situation can be partially
attributed to the knowledge gained from the analysis of the
SL-1 explosion. Other preventive measures of a thermal-
hydraulic nature that might be considered in view of present
knowledge are discussed in the following section.

B.   Preventive Factors

     Although numerous criteria must be considered when
assessing various nuclear reactor designs, the following
discussion is based on the thermal vapor explosion hazard
only, where an attempt is made to clarify the measures to
be considered to diminish the potential for explosive vaporization resulting from a severe hypothetical reactor core
meltdown event.

     Since intimate molten fuel contact with liquid coolant
appears to be an important criterion necessary for initiating
a vapor explosion, factors that tend to separate fuel from
coolant should diminish the potential for such an explosion.
The presence of noncondensable gases can lead to such separation, in addition to the effect of damping out pressurization.

plained by a film boiling destabilization temperature. In addition, pressure effects have been shown also to cause film boiling collapse. The recent work of Lienhard(*139*), Dhir(*140*), Gunnerson and Cronenberg(*41,43,138*), and Bankoff et al(*141*) with respect to understanding the conditions for film boiling stability should help in understanding vapor explosion criteria.

Several new approaches to the fragmentation process also have been developed. In a general sense, the violent boiling-collapse mechanisms appear promising, although significant questions remain concerning an adequate modeling of the collapse process and associated energy imparted to the hot material. In addition, most of this work has not considered the role of solidification on the breakup process; yet it has been shown that the molecular crystallization theory(*49,50*) would predict the commencement of surface solidification within times less than a millisecond for oxide fuels that undergo the large heat transfer quenching rates for good contact conditions. The work presented in References *82* and *83* attempts to couple such boiling concepts with surface solidification; thus, it appears to be a consistent approach. The mixing and vapor blanketing effects discussed by Cho(*43,46*) also deserve further consideration and should be incorporated into modeling efforts.

The fact remains, however, that an all-encompassing fragmentation mechanism has not been widely accepted; indeed, the situation of many conflicting model concepts exists. The problem is complicated by the fact that certain concepts (surface solidification, gas release, etc.) may describe a particular material and experimental system (UO_2 in Na, Ag in H_2,...), but may not be relevant to the vapor explosion problem. This is not to say that an understanding of fragmentation is unimportant; indeed it is, since it is the one necessary condition that is generally accepted. However, it appears desirable to first understand the overall nature of the phenomena involved in explosive vaporization and at that point, relate the details of fragmentation to the problem.

Besides some of the suggestions given in Reference *53*, further research efforts might include:

1.   injection or entrapment experiments to provide the conditions for triggering vapor explosions

2.   large-scale experiments to demonstrate
     the effects of geometry, size, mass
     ratio, and ambient pressure effects
     on the pressure detonation model

3.   shock-wave experiments to measure the
     number of fragments as a function of
     the pressure pulse and energy

4.   further theoretical development of
     fragmentation and intermixing models
     based on the pressure wave initiator
     concept

5.   clarification of the effect of transient
     solidification of the molten material
     during quenching on the fragmentation
     process and its influence on the spon
     taneous nucleation model

6.   experimental and theoretical studies on
     dynamic surface tension between suddenly
     contacting materials as applied to the
     spontaneous nucleation model

7.   experimental and theoretical studies to
     assess the temperature and pressure con-
     ditions for destabilization of film boiling

REFERENCES

1.   Witte, L. C. and Cox, J. E., "Thermal Explosion
     Hazards," Advances in Nuclear Science and Technology,
     Vol. 7, pp 329-364, 1973.

2.   Long, G. "Explosions of Molten Aluminum in Water:
     Cause and Prevention," Metal Progress, Vol. 71,
     pp 107-112, 1957.

3.   Epstein, L. F., "Recent Developments in the Study of
     Metal-Water Reactions," Progress in Nuclear Energy,
     Series IV, pp 461-483, 1961.

4.   Cronenberg, A. W. and Grolmes, M. A., "Fragmentation
     Modeling Relative to the Breakup of Molten UO_2 in
     Sodium," Journal of Nuclear Safety, Vol. 16, pp 683-
     700, 1975.

5.   Ivins, R. O. et al, "Reaction of Water as Initiated by
     a Power Excursion in a Nuclear Reactor (TREAT),"
     Nuclear Science and Engineering, Vol. 25, pp 131-140,
     1966.

6.   McLain, H. A., "Potential Metal-Water Reactions in
     Light-Water Cooled Power Reactors," USAEC Report No.
     ORNL-NSIC-23, August, 1968.

7.   Buxton, L. D. and Nelson, L. S., "Steam Explosions,"
     Sandia Laboratory Report No. 74-0382 on Core Meltdown
     Review, August, 1975.

8.   Reactor Safety Study, WASH-1400, U. S. Nuclear Regula-
     tory Commission, October, 1975.

9.   Rengstorff, G. W., Lemmon, A. W., Hoffman, A. H.,
     "Review of Knowledge of Explosions Between Aluminum and
     Water," Report to Aluminum Association, April 11, 1969.

10. Enger, T. and Hartman, D., "Rapid Phase Transformation
     During LNG Spillage on Water," Proceedings of the 3rd
     Conference on Liquified Natural Gas, Washington, D. C.,
     September, 1972.

11.   Katz, D. L. and Sliepcevich, C. M., "LNG/Water Explosions: Cause and Effect," Hydrocarbon Processing, pp 240-244, November, 1971.

12.   Hatfield, G. W., "A Reactor Emergency with Resulting Improvements," Mechanical Engineering, Vol. 77, pp 124-126, 1955.

13.   Hurst, D. G., "The Accident to the NRX Reactor," AECL-233, 1953.

14.   Zinn, W. H., A Letter on the EBR-1 Fuel Meltdown, Nucleonics, Vol. 14, 1955.

15.   Brittan, R. O., "Analysis of the EBR-I Core Meltdown," Proceedings of 2nd UN Conference on Peaceful Uses of Atomic Energy, Vol. 12, Geneva, 1958.

16.   Dietrich, J. R., "Experimental Determination of the Self-regulation and Safety of Operating Water-moderated Reactors," Proceedings of the International Conference on Peaceful Uses of Atomic Energy, Vol. 13, pp 88-101, 1955.

17.   Dietrich, J. R., "Experimental Investigation of the Self-limitation of Power Driving Reactivity Transients in a Subcooled, Water-moderated Reactor," ANL-5323, 1954.

18.   Thompson, T. J. and Beckerly, J. G., The Technology of Nuclear Reactor Safety 1, M. I. T. Press, Page 672, 1964.

19.   Miller, R. W., Sola, A. McCardell, R. K., "Report of the SPERT-I Destructive Test Program on an Aluminum Plate-type Water-moderated Reactor," IDO-16883, 1964.

20.   Higgins, H. M. and Schultz, R. D., "The Reaction of Metals with Water and Oxidizing Gases at High Temperatures," IDO-28000, 1957.

21.   Reactor Development Program Progress Report, ANL-6904, pp 102-104, May, 1964.

22.   Miller, R. W. et al, "Experimental Results and Damage Effects of Destructive Test," Transactions of the American Nuclear Society, Vol. 6, Page 138, 1963.

23. DiSalvo, R., "Phenomenological Investigation of Postu-
     lated Meltdown Accidents in Light-water Reactors,"
     Journal of Nuclear Safety, Vol. 18, pp 60-78, 1977.

24. Fauske, H. K., "CSNI Meeting on Fuel-Coolant Inter-
     actions," Journal of Nuclear Safety, Vol. 16,
     pp 436-443, 1975.

25. Tong, L. S. and Bennett, G. L., "NRC Water-Reactor
     Safety Research Program," Journal of Nuclear Safety,
     Vol. 18, pp 1-40, 1977.

26. Kelber, C. N., "Phenomenological Research in LMFBR
     Accident Analysis," Journal of Nuclear Safety, Vol. 14,
     1973.

27. Hicks, E. P. and Menzies, D. C., "Theoretical Studies
     on the Fast Reactor Maximum Accident," Proceedings,
     Conference on Safety, Fuels, and Core Design in Large
     Fast Power Reactor, ANL-7120, 1965.

28. Judd, A. M., "Calculation of the Thermodynamic Effic-
     iency of Molten Fuel-Coolant Interaction," Transactions
     of the American Nuclear Society, Vol. 13, Page 369, 1970.

29. Fauske, H. K., "On the Mechanisms of Uranium Dioxide-
     Sodium Explosive Interactions," Nuclear Science and
     Engineering, Vol. 51, pp 95-101, 1973.

30. Fauske, H. K., "The Role of Energetic Mixed Oxide Fuel-
     Sodium Thermal Interactions in LMFBR Safety," Proceed-
     ings of the 3rd CSNI Specialist Meeting on Na-Fuel
     Interactions in Fast Reactors, Tokyo, Japan, March, 1976.

31. Armstrong, D. R., Testa, F. T., Raridon, D. C., "Inter-
     action of Sodium with Molten UO_2 and Stainless Steel
     Using a Dropping Mode of Contact, ANL-7890, December,
     1971.

32. Armstrong, D. R., Goldfuss, G. T., Gebner, R. H.,
     "Explosive Interaction of Molten UO_2 and Liquid Sodium,"
     ANL-76-24, March, 1976.

33. Anderson, R. P. and Armstrong, D. R., "R-22 Vapor Ex-
     plosions," Annual ASME Winter Meeting: Nuclear Reactor
     Safety Heat Transfer Section, Atlanta, Georgia,
     November 27-December 2, 1977, pp 31-45.

34. Board, S. J., Hall, R. W., Brown, G. E., "The Rate of
    Spontaneous Nucleation in Thermal Explosion:
    Freon/Water Experiments," RD/BIN-3007, June, 1974.

35. Henry, R. E., Fauske, H. K., McUmber, L. M., "Vapor
    Explosion Experiments with Simulant Fluids," Proceed-
    ings of the ANS Conference on Fast Reactor Safety,
    Chicago, Illinois, October, 1976.

36. Henry. R. E. and McUmber, L. M., "Vapor Explosive
    Behavior at Elevated Ambient Pressure," ANL-77-34,
    pp 113-120, 1976.

37. Henry, R. E. and Fauske, H. K., "Nucleation Character-
    istics in Physical Explosion," Proceedings of 3rd
    Specialist Meeting on Na/Fuel Interaction in Fast
    Reactors, Tokyo, Japan, March 22-26, 1976, pp 596-623.

38. Board, S. J. and Hall, R. W., "Recent Advances in
    Understanding Large-scale Vapor Explosions," Proceedings
    of 3rd Specialist Meeting on Na/Fuel Interactions in Fast
    Reactors, Tokyo, Japan, March 22-26, 1976, pp 249-284.

39. Spiegler, P., et al, "Onset of Stable Film Boiling and
    the Foam Limit," International Journal of Heat and Mass
    Transfer, Vol. 6, pp 987-994, 1963.

40. Gunnerson, F. S. and Cronenberg, A. W., "A Correlation
    for the Leidenfrost Temperature for Spherical Particles
    and Its Application to FCI Analysis," Transactions of
    the American Nuclear Society, Vol. 25, pp 381-383, o977.

41. Henry, R. E., "A Correlation for the Minimum Film Boil-
    ing Temperature," 14th National Heat Transfer Conference,
    AICHE-ASME, Atlanta, Georgia, 1973.

42. Gunnerson, F. S. and Cronenberg, A. W., "On the Thermo-
    dynamic Superheat Limit for Liquid Metals and Its
    Relation to the Leidenfrost Temperature," (submitted
    to Journal of Heat Transfer, 1978).

43. Waldram, K. L., "Interaction of Low Boiling Point Liquid
    Drops on a Hot Liquid Surface," M. S. Thesis, North-
    western University, Evanston, Illinois, April, 1974.

44. Hall, W. B., "Bubble Growth with Acoustic Loading," (Calculations presented at OECD CSNI Meeting, Argonne National Laboratory, Argonne, Illinois, December 8-9, 1977).

45. Cho, D. H., Fauske, H. K., Grolmes, M. A., "Some Aspects of Mixing in Large-Mass, Energetic Fuel-Coolant Interactions," Proceedings of the ANS Conference on Fast Reactor Safety, Chicago, Illinois, October, 1976.

46. Henry, R. E. and Cho, D. H., "An Evaluation of the Potential for Energetic Fuel-Coolant Interactions in Hypothetical LMFBR Accidents," Annual ASME Winter Meeting, Nuclear Reactor Safety Heat Transfer Section, Atlanta, Georgia, November 27-December 2, 1977, pp223-237.

47. Chalmers, B., Principles of Solidification, John Wiley Publishing Company, New York, 1964.

48. Turnbull, D., "Formation of Crystal Nuclei in Liquid Metals," Journal of Applied Physics, Vol. 21, pp 1022-1028, 1950.

49. Cronenberg, A. W. and Fauske, H. K., "UO Solidification Associated with Rapid Quenching in Liquid Sodium," Journal of Nuclear Materials, Vol. 52, pp 24-32, 1974.

50. Cronenberg, A. W. and Coats, R. L., "Solidification Phenomena for UO_2, UC, and UN Relative to Quenching in Na Coolant," Nuclear Engineering and Design, Vol. 36, pp 261-272, 1976.

51. Ladisch, R., "Comment on Fragmentation of UO_2 by Thermal Stress and Pressurization," Nuclear Engineering and Design, Vol. 43, pp 327-328, 1977.

52. Board, S. J., Hall, R. W. and Hall, R. S., "Detonation of Fuel Coolant Explosions," Nature, Vol. 254, pp 319-321, March, 1975.

53. Board, S. J. and Caldarola, L., "Fuel-Coolant Interaction in Fast Reactors," Annual ASME Winter Meeting, Nuclear Reactor Safety Heat Transfer Section, Atlanta, Georgia, November 27-December 2, 1977, pp 195-222.

54. Simpkins, P. G. and Bales, E. L., "Water-drop Response
     to Sudden Accelerations," Journal of Fluid Mechanics,
     Vol. 55, Page 629, 1972.

55. Bankoff, S. G. and Jo, J. H., "On the Existence of
     Steady-State Fuel Coolant Thermal Detonation Waves,"
     Northwestern University Report NU-2512-8, 1976.

56. Bankoff, S. G., "Vapor Explosion: A Critical Review,"
     Proceedings of the 6th International Heat Transfer
     Conference, Toronto, Canada, August, 1978.

57. Patel, P. D. and Theofanous, T. G., "Fragmentation
     Requirements for Detonating Vapor Explosions," Purdue
     University Report PNE-78-122, 1978.

58. Williams, D. C., "A Critique of the Board-Hall Model
     for Thermal Detonations in UO_2-Na Systems," Proceedings
     of the ANS Conference on Fast Reactor Safety, Chicago,
     Illinois, October, 1976.

59. Sharon, A. and Bankoff, S. G., "Propagation of Shock
     Waves Through a Fuel/Coolant Mixture; Part A: Boundary
     Layer Stripping Mechanism," Northwestern University
     Report COO-2512-12, March, 1978.

60. Ivins, R. O., "Interactions of Fuel, Cladding, and
     Coolant, ANL-7399, pp 162-165, 1967.

61. Hinze, J. O., "Forced Deformations of Viscous Liquid
     Globules," Applied Scientific Research (A), 1,
     pp 263-272, 1949.

62. Cho, D. H. and Gunther, W. H., "Fragmentation of Molten
     Materials Dropped into Water," Transactions of the
     American Nuclear Society, Vol. 16, pp 185-186, 1973.

63. Lazarrus, J., Navarre, J. P., Kottowski, H. M., "Thermal
     Interaction Experiments in a Channel Geometry Using
     Al_2O_3 and Na," Proceedings of 2nd CSNI Specialist Meet-
     ing on Na-Fuel Interactions in Fast Reactors, Ispra,
     Italy, November, 1973.

64. Paoli, R. M. and Mesler, R. B., "Explosion of Molten
     Lead in Water," Proceedings of the Conference on High-
     speed Photography, Stockholm, Sweden, 1968.

65. Darby, K. et al, "The Thermal Interaction Between
    Water and Molten Aluminum under Impact Conditions in
    a Strong Tube," Proceedings of the International Con-
    ference on Fast Reactors for Safe and Reliable
    Operation, Karlsruhe, Germany, October, 1972.

66. Board, S. J., Farmer, C. L., Poole, D. H., "Fragmen-
    tation in Thermal Explosions," International Journal
    of Heat and Mass Transfer, Vol. 17, pp 331-339, 1974.

67. Flory, K., Paoli, R. M., Mesler, R. B., "Molten Metal-
    Water Explosions," Chemical Engineering Progress, Vol.
    65, pp 50-54, December, 1969.

68. Swift, D. and Baker, L., Reactor Development Program
    Report, ANL-7152, pp 87-96, January, 1965.

69. Mizuta, H., "Fragmentation of Uranium Dioxide after
    Molten UO_2-Na Interaction," Nuclear Science and
    Technology, Vol. 11, pp 480-487, 1974.

70. Martinson, Z. R., "Behavior of 5-inch long, 1/4 OD
    Zircaloy-2 Oxide Fuel Rods Subjected to High Energy
    Power Bursts, IN-ITR-107, August, 1969.

71. Wright, R. W. and Humberstone, G. H., "Dispersal and
    Pressure Generation by Water Impact upon Molten Alum-
    inum," Transactions of the American Nuclear Society,
    Vol. 9, Page 305, 1966.

72. Coffield, R. D. and Wattelet, P. L., "An Analytical
    Evaluation of Fuel Failure Propagation for the Fast
    Flux Test Facility," Report of AEC Contract AT(45-1)-2171,
    Task No. 1, WARD, Hanford Engineering Laboratory,
    November, 1970.

73. Bradley, R. H. and Witte, L. C., "Explosive Interaction
    of Molten Metals Injected into Water," Nuclear Science
    and Engineering, Vol. 48, pp 387-396, 1972.

74. Colgate, S. A. and Sigurgeirsson, T., "Dynamic Mixing
    of Water and Lava," Nature, Vol. 244, pp 552-555,
    August, 1973.

75. Roberts, K. V., "Theoretical Calculations of Fuel-
     Coolant Interactions," CREST Specialist Meeting on
     Sodium Fuel Interactions, Grenoble, France, January,
     1972.

76. Bruckner, U. and Unger, H., "Analyses Physiklalischer
     Vorgange bei Thermodynamischen Mischreaktionen,"
     Universität Stuttgart, Institut fur Kernenergetik,
     IKE-Bericht, No. 6-69, 1973.

77. Buchanan, D. J., "A Model for Fuel-Coolant Interactions,"
     Journal of Physics, D-7, pp 1441-1457, 1974.

78. Buchanan, D. J. and Dullforce, T. A., "Mechanism for
     Vapour Expansions," Nature, Vol. 245, pp 32-34,
     September, 1973.

79. Caldarola, L. and Kastenberg, W. E., "On the Mechanism
     of Fragmentation during Molten Fuel-Coolant Thermal
     Interactions," Proceedings of American Nuclear Society
     Conference on Fast Reactor Safety, Los Angeles, 1974.

80. Vaughan, G. J., Caldarola, L., Todreas, N. E., "A Model
     for Fuel Fragmentation during Molten Fuel/Coolant Ther-
     mal Interactions," Proceedings of the American Nuclear
     Society Conference on Fast Reactor Safety, Chicago,
     Illinois, October, 1976.

81. Stevens, J. W. and Witte, L. C., "Destabilization of
     Vapor Film Boiling Around Spheres," International
     Journal of Heat and Mass Transfer, Vol. 16, pp 669-
     678, 1973.

82. Benz, R., Frohlich, G., Unger, H., "Fragmentationsver-
     lauf heisser flussiger Schmelze in Wasser, beschrieben
     mit einem Dampfblasenkollapsmodell," Universität
     Stuttgart, Institut für Kernenergetik, Reaktortagung,
     Düsseldorf, 1976.

83. Benz, R. et al, "Theoretische Studien zur Damfexplosion,"
     and (2), "Technischer Fachbericht zum Forschungsvorhaben,"
     BMFT RS76, April, 1976.

84. Benz, R., Fröhlich, G. and Unger, H., "Ein Dampfblasen-
     kollapsmodell (DBK-Modell) zur Beschreibung des Fragmen-
     tationsverlaufs heisser, flussiger Schmelze in Wasser,"
     Atomkernenergie, Vol. 29, pp 261-265, 1977.

85. Bevis, M. K. and Fielding, P. J., "Numerical Solution of Incompressible Bubble Collapse with Jetting," in Moving Boundary Problem in Heat Flow and Diffusion, J. R. Ockendon and W. R. Hodgkins, England, Clarendon Pewaa, 1975.

86. Christiansen, J. P., "Numerical Simulation of Hydrodynamics by the Method of Point Vortices," Journal of Computational Physics, Vol. 13, pp 363-379, 1973.

87. Dullforce, T. A., Buchanan, D. J., Peckover, R. S., "Self-triggering and Small-scale Fuel-Coolant Interactions: Experiment," Journal of Physical Dynamics, Vol. 9, pp 1295-1303, 1976.

88. Bjorkquist, G. M., "An Experimental Investigation of the Fragmentation of Molten Metals in Water," TID-26826, 1975.

89. Benjamin, T. B. and Ellis, A. T., "The Collapse of Cavitation Bubbles and the Pressure Thereby Produced Against Solid Boundaries," Philosophical Transactions of the Royal Society of London (A), Page 221, 1966.

90. Peckover, R. S., Buchanan, D. J., Ashby, D. E. T. F., "Fuel-Coolant Interactions in Submarine Volcanism," Nature, Vol. 245, pp 307-308, October, 1973.

91. Caldarola, L., "A Theoretical Model for Molten Fuel-Sodium Interaction on a Nuclear Fast Reactor," Nuclear Engineering and Design, Vol. 22, pp 175-211, 1972.

92. Caldarola, L., "A Theoretical Model with Variable Masses for the Molten Fuel-Sodium Thermal Interaction in a Nuclear Reactor," Nuclear Engineering and Design, Vol. 34, pp 181-201, 1975.

93. Schlechtendahl, S., "Sieden des Kuhlmittels in Natrium Gekuhlten Schnellen Reaktoven," Karlsruhe Nuclear Research Center Report, KFK 1020, June, 1969.

94. Farahat, M. M., "Transient-Boiling Heat Transfer from Spheres to Sodium," ANL-7909, January, 1972.

95. Farahat, M. M., Armstrong, D. R., Eggen, D. T.,
     "Pool Boiling in Subcooled Sodium at Atmospheric
     Pressure," Nuclear Science and Engineering, Vol. 53,
     pp 240-253, 1974.

96. Zyszkowski, W., "Experimental Investigation of Fuel-
     Coolant Interaction," Nuclear Technology, Vol. 33,
     pp 40-59, April, 1977.

97. Fröhlich, G., Schmidt, E., Osswald, H., "Dampfexplos-
     ionen bei Thermischen Reaktionen von zwei Flüssigkeiten,
     IKE Report K-44, July, 1973.

98. Shiralkar, G. S. and Todreas, N. E., "An Investigation
     of Fragmentation of Molten Metals Dropped into Cold
     Water," MIT Report No. COO-2781-7TR, November, 1976.

99. Witte, L. C. et al, "Heat Transfer and Fragmentation
     During Molten-Metal/Water Interactions," Journal of
     Heat Transfer, Vol. 95, pp 521-527, November, 1973.

100. Plesset, M. S. and Chapman, R. B., "Collapse of an
     Initially Spherical Vapor Cavity in the Neighborhood
     of a Solid Boundary," Journal of Fluid Mechanics,
     Vol. 47, pp 283-290, 1970.

101. Beer, H., "Beitrag zur Warmeubertragung beim Sieden,"
     Progress in Heat and Mass Transfer, Vol. 2, Oxford,
     Pergamon Press, pp 311-370, 1969.

102. Fröhlich, G., Müller, G. and Unger, H., "Experiments
     with Water and Hot Melts of Lead," Journal of Non-
     Equilibrium Thermodynamics, Vol. 1, pp 91-103, 1976.

103. Zyszkowski, W., "On the Transplosion Phenomena and the
     Leidenfrost Temperature for Molten Copper-Water Thermal
     Interaction," International Journal of Heat and Mass
     Transfer, Vol. 19, pp 623-625, 1976.

104. Hess, P. D. and Brondyke, K. J., "Causes of Molten
     Aluminum-Water Explosions and their Prevention," Metal
     Progress, Vol. 95, pp 93-100, April, 1969.

105. Sallack, J. A., "On Investigation of Explosions in the Soda Smelt Dissolving Operation," Pulp Paper Magazine of Canada, 56, pp 114-118, 1955.

106. Brauer, F. E., Green, N. W., Mesler, R. B., "Metal/Water Explosions," Nuclear Science and Engineering, Vol. 31, pp 551-554, 1968.

107. Schins, H., "The Consistent Boiling Model for Fragmentation in Mild Thermal Interaction-Boundary Conditions," WURATOM Report, EUR/c-IS/699/73e, 1973.

108. Kazimi, M. S., "Theoretical Studies of Some Aspects of Molten Fuel-Coolant Thermal Interactions," Science Doctorate Thesis, MIT, Cambridge, Massachusetts, May, 1973.

109. Bernath, L., "Theory of Bubble Formation in Liquids," Industrial and Engineering Chemistry, Vol. 44, pp 1310-1313, 1951.

110. Cronenberg, A. W. and Grolmes, M. A., "A Review of Fragmentation Models Relative to Molten UO_2 Breakup When Quenched in Sodium Coolant," ASME Paper 74-WA/HT-42, 1974.

111. Flynn, H. G., "Physics of Acoustic Cavitation in Liquids," Physical Acoustics, Vol. 1 (B), New York, W. P. Mason, Academic Press, 1964.

112. Epstein, M., "A New Look at the Cause of Thermal Fragmentation," Transactions of the American Nuclear Society, Vol. 19, Page 249, 1974.

113. Epstein, M., "Thermal Fragmentation--A Gas Release Phenomenon," Nuclear Science and Engineering, Vol. 55, pp 462-467, 1974.

114. Buxton, L. D. and Nelson, L. S., "Impulse Initiated Gas Release--A Possible Trigger for Vapor Explosions," Transactions of the American Nuclear Society, Vol. 26, Page 398, 1977.

115. Nelson, L. S. and Buxton, L. D., "The Thermal Interaction of Molten LWR Core Materials with Water," Transactions of the American Nuclear Society, Vol. 26, Page 397, 1977.

116.   Fast, J. D., "Interaction of Metals and Gases,"
       Thermodynamics and Phase Relations, Vol. 1, New York,
       Academic Press, 1965.

117.   Johnson, G. W. and Shuttleworth, R., "The Solubility
       of Krypton in Liquid Lead, Tin and Silver," Philo-
       sophical Magazine, Vol. 4, Page 957, 1959.

118.   Gunnerson, F. S. and Cronenberg, A. W., "A Prediction
       of the Inert Gas Solubilities in Stoichiometric Molten
       UO_2," Journal of Nuclear Materials, Vol. 58, pp 311-320,
       1975.

119.   McLain, H. A., "Potential Metal-Water Reactions in
       Light-Water-Cooled Power Reactors, ORNL-NSIC-233,
       August, 1968.

120.   Zyszkowski, "Experimental Investigation of Fuel-Coolant
       Interaction," Nuclear Technology, Vol. 33, pp 40-59,
       April, 1977.

121.   Hsiao, K. H. et al, "Pressurization of a Solidifying
       Sphere," Journal of Applied Mechanics, Vol. 39,
       pp 71-77, 1972.

122.   Cronenberg, A. W., Chawla, T. C., Fauske, H. K., "A
       Thermal Stress Mechanism for the Fragmentation of
       Molten UO_2 Upon Contact with Sodium Coolant," Nuclear
       Engineering and Design, Vol. 30, pp 443-454, 1974.

123.   Knapp, R. B. and Todreas, N. E., "Thermal Stress
       Initiated Fracture as a Fragmentation Mechanism in
       UO_2-Na Fuel-Coolant Interaction," Nuclear Engineering
       and Design, Vol. 35, pp 69-76, 1975.

124.   Wright, R. W. and Humberstone, G. H., "Dispersal and
       Pressure Generation by Water Impact upon Molten Alumin-
       um," Transactions of American Nuclear Society, Vol. 9,
       Page 305, 1966.

125.   Zyszkowski, W., "Thermal Interaction of Molten Copper
       with Water," International Journal of Heat and Mass
       Transfer, Vol. 18, pp 271-287, 1975.

126.   SL-1 Project, Final Report of the SL-1 Recovery Oper-
       ation, IDO-19311, 1962.

127. Boudreau, J. E. and Jackson, J. F., "Recriticality
      Considerations in LMFBR Accidents," Proceedings of
      the ANS Conference on Fast Reactor Safety, Beverly
      Hills, California, April, 1974, pp 1265-1289.

128. Wright, R. W. et al, "Summary of Autoclave TREAT Tests
      on Molten Fuel Coolant Interactions," Proceedings of
      the American Nuclear Society Conference on Fast Reactor
      Safety, Beverly Hills, California, April, 1976,
      pp 254-267.

129. Epstein, M. and Cho, D. H., "Fuel Vaporization and
      Quenching by Cold Sodium; Interpretation of TREAT Test
      S-11," Proceedings of the American Nuclear Society
      Conference on Fast Reactor Safety, Beverly Hills,
      California, April, 1976, pp 268-278.

130. Dickerman, C. E., "U. S. Studies on LMFBR Fuel Be-
      havior under Accident Conditions," Journal of Nuclear
      Safety, Vol. 14, pp 452-460, 1973.

131. Cronenberg, A. W. and Grolmes, M. A., "An Assessment
      of the Coolant Voiding Dynamics Following the Failure
      of Preirradiated LMFBR Fuel Pins," Nuclear Technology,
      Vol. 27, pp 395-410, 1975.

132. Cronenberg, A. W., "A Thermohydrodynamic Model for
      Molten UO_2-Na Interactions, Pertaining to Fast Reactor
      Fuel Failure Accidents," ANL-7947, June, 1972.

133. Cronenberg, A. W., Fauske, H. K., Eggen, D. T.,
      "Analysis of the Coolant Behavior Following Fuel
      Failure and Molten Fuel-Sodium Interaction in a Fast
      Nuclear Reactor," Nuclear Science and Engineering,
      Vol. 50, pp 53-62, 1973.

134. Amblard, M. et al, "Out-of-Pile Studies in France on
      Sodium-Fuel Interaction," Proceedings of the American
      Nuclear Society Conference on Fast Reactor Safety,
      Beverly Hills, California, April 2-4, pp 910-921, 1974.

135. Amblard, M., "Preliminary Results of a Contact Between
      4Kg of Molten UO_2 and Liquid Sodium," Proceedings of
      the 3rd Specialist Meeting on Na/Fuel Interaction in
      Fast Reactors, Tokyo, Japan, March 26-27, 1976,
      pp 545-560.

136. Cho, D. H., Ivins, R. O., Wright, R. W., "A Rate-
      Limited Model of Molten-Fuel/Coolant Interactions:
      Model Development and Preliminary Calculations,
      ANL-7919, March, 1972.

137. Wright, R. W. and Cho, D. H., "Acoustic and Inertial
      Constraints in Molten Fuel-Coolant Interactions,"
      Transactions of the American Nuclear Society, Vol. 13,
      Page 658, 1970.

138. Gunnerson, F. S. and Cronenberg, A. W., "A Thermodynamic
      Prediction of the Temperature for Film Boiling De-
      stabilization and Its Relation to Vapor Explosion
      Phenomena," Transactions of the American Nuclear Society,
      1978.

139. Leinhard, J. H. and Wong, P. T. Y., "The Dominant and
      Unstable Wavelength and Minimum Heat Flux during Film
      Boiling on a Horizontal Cylinder," Journal of Heat
      Transfer, Vol. 86, pp 220-226, May, 1964.

140. Dhir, V. K. and Purohit, G. P., "Subcooled Film-
      boiling Heat Transfer from Sphere," ASME Paper
      77-HT-78, 1977.

141. Bankoff, S. G. et al, "Destabilization of Film Boiling
      in Liquid-Liquid Systems," Proceedings of the 6th
      International Heat Transfer Conference, Toronto,
      Canada, August, 1978.